NIST Technical Note 1727

Annex 47 Report 3: Commissioning Cost-Benefit and Persistence of Savings

Hannah Friedman
Marti Frank
Kristin Heinemeier
Kim Crossman
Eliot Crowe
Portland Energy Conservation Inc.

David Claridge
Cory Toole
Texas A&M University

Daniel Choinière
Natural Resources Canada
CanmetENERGY
Energy Technology Centre

Natascha Milesi Ferretti
National Institute of Standards and Technology
Engineering Laboratory

http://dx.doi.org/10.6028/NIST.TN.1727

December 2011

U.S. Department of Commerce
Rebecca Blank, Acting Secretary

National Institute of Standards and Technology
Patrick D. Gallagher, Under Secretary of Commerce for Standards and Technology and Director

CITATION

Friedman, H. et al. "Annex 47 Report 3: Commissioning Cost-Benefit and Persistence of Savings", National Institute of Standards and Technology, Technical Note 1727. December 2011.

This publication was simultaneously published by Natural Resources Canada as Friedman, H. et al. "Annex 47 Report 3: Commissioning Cost-Benefit and Persistence of Savings", a Report of Cost-Effective Commissioning of Existing and Low Energy Buildings. December 2011.

Copies of this report may be obtained from the Annex 47 web site at: http://www.iea-annex47.org , the NIST website at: http://www.nist.gov, or from the IEA/ECBCS Bookshop at: www.ecbcs.org.

International Energy Agency
Energy Conservation in
Buildings and Community
Systems Programme

Preface

International Energy Agency

The International Energy Agency (IEA) was established in 1974 within the framework of the Organization for Economic Co-operation and Development (OECD) to implement an international energy program. A basic aim of the IEA is to foster co-operation among the twenty-eight IEA participating countries and to increase energy security through energy conservation, development of alternative energy sources and energy research, development and demonstration (RD&D).

Energy Conservation in Buildings and Community Systems

The IEA coordinates research and development in a number of areas related to energy. The mission of one of those areas, the Energy Conservation for Building and Community Systems Program (ECBCS) is to develop and facilitate the integration of technologies and processes for energy efficiency and conservation into healthy, low emission, and sustainable buildings and communities, through innovation and research.

The research and development strategies of the ECBCS Program [1] are derived from research drivers, national programs within IEA countries, and the IEA Future Building Forum Think Tank Workshop, held in March 2007. The R&D strategies represent a collective input of the Executive Committee members to exploit technological opportunities to save energy in the buildings sector, and to remove technical obstacles to market penetration of new energy conservation technologies. The R&D strategies apply to residential, commercial, office buildings and community systems, and will impact the building industry in three focus areas of R&D activities:

- Dissemination;
- Decision-making; and
- Building products and systems.

Participating countries in ECBCS: Australia, Austria, Belgium, Canada, P.R. China, Czech Republic, Denmark, Finland, France, Germany, Greece, Italy, Japan, Republic of Korea, the Netherlands, New Zealand, Norway, Poland, Portugal, Spain, Sweden, Switzerland, Turkey, United Kingdom and the United States of America.

The Executive Committee

Overall control of the program is maintained by an Executive Committee, which not only monitors existing projects but also identifies new areas where collaborative effort may be beneficial. To date the following projects have been initiated by the executive committee on Energy Conservation in Buildings and Community Systems (completed projects are identified by (*)):

Annex 1:   Load Energy Determination of Buildings (*)

Annex 2:   Ekistics and Advanced Community Energy Systems (*)

Annex 3:   Energy Conservation in Residential Buildings (*)

Annex 4:   Glasgow Commercial Building Monitoring (*)

Annex 5:   Air Infiltration and Ventilation Centre

Annex 6:   Energy Systems and Design of Communities (*)

Annex 7:   Local Government Energy Planning (*)

Annex 8:   Inhabitants Behaviour with Regard to Ventilation (*)

Annex 9:   Minimum Ventilation Rates (*)

Annex 10:   Building HVAC System Simulation (*)

Annex 11:   Energy Auditing (*)

Annex 12:   Windows and Fenestration (*)

Annex 13:   Energy Management in Hospitals (*)

Annex 14:   Condensation and Energy (*)

Annex 15:   Energy Efficiency in Schools (*)

Annex 16:   BEMS 1- User Interfaces and System Integration (*)

Annex 17:   BEMS 2- Evaluation and Emulation Techniques (*)

Annex 18:   Demand Controlled Ventilation Systems (*)

Annex 19:   Low Slope Roof Systems (*)

Annex 20:   Air Flow Patterns within Buildings (*)

Annex 21:   Thermal Modelling (*)

Annex 22:   Energy Efficient Communities (*)

Annex 23:   Multi Zone Air Flow Modelling (COMIS) (*)

Annex 24:   Heat, Air and Moisture Transfer in Envelopes (*)

Annex 25:   Real-time HVAC Simulation (*)

Annex 26:   Energy Efficient Ventilation of Large Enclosures (*)

Annex 27:   Evaluation and Demonstration of Domestic Ventilation Systems (*)

Annex 47

The objectives of Annex 47 were to enable the effective commissioning of existing and future buildings in order to improve their operating performance and to advance the state-of-the-art of building commissioning by:

Extending previously developed methods and tools to address advanced systems and low energy buildings, utilizing design data and the buildings' own systems in commissioning;

Automating the commissioning process to the extent practicable;

Developing methodologies and tools to improve operation of buildings in use, including identifying the best energy saving opportunities in HVAC system renovations; and

Quantifying and improving the costs and benefits of commissioning, including the persistence of benefits and the role of automated tools in improving persistence and reducing costs without sacrificing other important commissioning considerations.

To accomplish these objectives Annex 47 has conducted research and development in the framework of the following three areas:

Initial Commissioning of Advanced and Low Energy Building Systems

This area addressed what can be done for (the design of) future buildings to enable cost-effective commissioning. The focus was set on the concept, design, construction, acceptance, and early operation phase of buildings.

Commissioning and Optimization of Existing Buildings

This area addressed needs for existing buildings and systems to conduct cost-effective commissioning. The focus here was set on existing buildings where the commissioning process must be performed with incomplete or out-of-date documentation.

Commissioning Cost-Benefits and Persistence

This area addressed how the cost-benefit situation can be represented. Key answers were provided by developing international consensus methods for evaluating commissioning cost-benefit and persistence. The methods were implemented in a cost-benefit and persistence database using field data.

Annex 47 was an international joint effort conducted by 50 organizations in 11 countries:

Belgium	• KaHo St-Lieven • Ghent University • Passive House Platform • Université de Liège • Katholieke Universiteit Leuven
Canada	• Natural Resources Canada (CETC-Varennes) • Public Works and Governmental Services Canada • Palais de Congres de Montreal • Hydro Quebec
Czech Republic	• Czech Technical University
Finland	• VTT Technical Research Centre of Finland • Helsinki University of Technology
Germany	• Ebert-Baumann Consulting Engineers • Institute of Building Services and Energy Design • Fraunhofer Institute for Solar Energy Systems
Hong Kong/China	• Hong Kong Polytechnic University
Hungary	• University of Pécs
Japan	• Kyoto University • Kyushu University • Chubu University • Okayama University of Science • NTT Facilities • Osaka Gas Co. • Kansai Electric Power Co. • Kyushu Electric Power Co. • SANKO Air Conditioning Co • Daikin Air-conditioning and Environmental Lab • Tokyo Electric Power Co • Tokyo Gas Co. • Takenaka Corp. • Chubu Electric Power Co. • Tokyo Gas Co., Ltd. • Tonets Corp • Nikken Sekkei Ltd • Hitachi Plant Technologies • Mori Building Co. • Takasago Thermal Engineering Co., Ltd. • Institute for Building Environment and Energy Conservation
Netherlands	• TNO Environment and Geosciences • University of Delft
Norway	• Norwegian University of Science and Technology • SINTEF
USA	• National Institute of Standards and Technology • Texas A&M University • Portland Energy Conservation Inc. • Carnegie Mellon University • Johnson Controls • Siemens • Lawrence Berkeley National Laboratory

FOREWORD

This report summarizes part of the work of IEA-ECBCS Annex 47 **Cost-Effective Commissioning of Existing and Low Energy Buildings**. It is based on the research findings from the participating countries. The publication is an official Annex 47 report.

Report 1, 'Commissioning Overview' can be considered as an introduction to the commissioning process.

Report 2, 'Commissioning Tools for Existing and Low Energy Buildings' provides general information on the use of tools to enhance the commissioning of low energy and existing buildings, summarizes the specifications for tools developed in the Annex and presents building case studies.

Report 3, 'Commissioning Cost Benefit and Persistence' presents a collection of data that would be of use in promoting commissioning of new and existing buildings and defines methods for determining costs, benefits, and persistence of commissioning, The report also highlights national differences in the definition of commissioning.

Report 4, 'Flowcharts and Data Models for Initial Commissioning of Advanced and Low Energy Building Systems' provides a state of the art description of the use of flow charts and data models in the practice and research of initial commissioning of advanced and low energy building systems.

In many countries, commissioning is still an emerging activity and in all countries, advances are needed for greater formalization and standardization. We hope that this report will be useful to promote best practices, to advance its development and to serve as the basis of further research in this growing field.

Natascha Milesi Ferretti and Daniel Choinière

Annex 47 Co-Operating Agents

ACKNOWLEDGEMENT

The material presented in this publication has been collected and developed within an Annex of the IEA implementing agreement on Energy Conservation in Buildings and Community Systems, Annex 47, "**Cost-Effective Commissioning of Existing and Low Energy Buildings**".

This report, together with three companion Annex reports are the result of an international joint effort conducted in ten countries. All those who have contributed to the project are gratefully acknowledged.

On behalf of all participants, the members of the Executive Committee of IEA Energy Conservation in Building and Community Systems Implementing Agreement as well as the funding bodies are also gratefully acknowledged.

A list of participating countries, institutes, and people can be found at the end of this report.

REPORT EDITORS:

Hannah Friedman (Portland Energy Conservation, Inc)
David Claridge (Texas A&M University)
Daniel Choinière (Natural Resources Canada)
Natascha Milesi Ferretti (NIST, USA)

TABLE OF CONTENTS

EXECUTIVE SUMMARY

Background

Commissioning (Cx) of new and existing buildings has been shown to reduce energy usage and can also produce non-energy related benefits such as improved occupant comfort. When compared with other initiatives such as installation of high efficiency equipment or installing photovoltaic systems, commissioning is highly cost-effective, resulting in short investment payback periods.

Despite the proven benefits, commissioning is not "business as usual," and this is generally attributed to the following factors:
1. The costs and benefits of commissioning are not clearly understood by the decision-makers in the commercial buildings industry.
2. This (Point 1) is further complicated due to the lack of a single definition of what the commissioning process includes.
3. Commissioning focuses on the operation of systems and their interactions, and there is a perception that operational improvements may not persist over time.
4. While commissioning can be explained as a logical sequence of steps, the details are complex and the outcomes aren't tangible in the same way that, for example, high efficiency lighting is. That means it takes time for the buildings industry to become familiar and comfortable with the process.

This report is the result of an international research project that collected data to help overcome the barriers listed in items 1, 2, and 3.

The core purpose of this report was to collect data that would be of use in promoting commissioning of new and existing buildings. A secondary purpose was to define methods for determining costs, benefits, and persistence of commissioning along with understanding national differences in the definition of commissioning.

Research was grouped under two broad headings: *Commissioning Cost-Benefit*, and *Commissioning Persistence*.

Commissioning Cost-Benefit

Literature Review of Commissioning Cost-Benefit Methodologies

Twelve studies were summarized, focused on studies where the cost-benefit methodologies were known. The majority were research studies of multiple buildings, and the studies ranged from research reports, databases, and marketing literature. These studies are summarized in three main aspects:

1. Author, year, format, audience, caveats and considerations,
2. Quantitative data – number of buildings, costs, benefits, payback, and
3. Methodologies used to determine quantitative data – subdivided into simple, moderate and complex.

Creation of an International Cost-Benefit Database

Financial and technical data was collected and analyzed from ten new building commissioning projects and 44 existing building commissioning projects, from seven countries.

This subtask commenced with efforts to determine what key data was to be collected (divided into "required" and "optional" data), and to develop the data collection forms. These data to be collected included:

1. Description of what was included in the commissioning process,
2. Commissioning cost,
3. Energy and cost savings from commissioning,

4. What problems were found, and the solutions, and

5. Non-energy benefits.

Collected data was collated in spreadsheets for analysis and generation of charts of the key findings.

Commissioning Persistence

Literature Review on the Persistence of Commissioning Benefits

This review summarized the findings from five studies encompassing 37 commissioning projects from across the USA. Persistence of savings was expressed as a percentage of the original claimed savings, after a specified time has elapsed after the project (e.g., 75 % after five years). In addition to evaluating project savings, the studies covered persistence at the level of specific measures, including the reasons for measures not persisting.

Impact of Savings Normalization Method on Commissioning Persistence

This study reviewed two weather normalization methods that are used in calculating energy savings from commissioning, and compared their impact on commissioning persistence claims. The two methods evaluated were:

International Performance Measurement and Verification Protocol (IPMVP)

A baseline regression model (or calibrated simulation model) is created based on the pre-commissioning energy use and recorded temperature/humidity. In the post-commissioning period, weather data is collected, and the regression model is used to predict what the energy use would have been if commissioning hadn't occurred. The actual energy use is subtracted from the modeled prediction, and this constitutes the energy saved.

Normalized Annual Consumption (NAC)

Similar basic principle to the IPMVP; a regression model (or calibrated simulation model) is created using baseline data. This model is applied to a standardized 'average' weather year based on the site location in order to calculate baseline annual energy use. In the post-commissioning period the regression model is recreated using post-energy and post-weather data, and this regression model is applied to the same 'average' weather year. The difference between the two modeled average years constitutes the savings.

Examples of Tools for Enhancing Persistence of Commissioning Benefits

There are a number of data collection and analysis tools that may be used for monitoring the persistence of commissioning improvements. This study described two such tools:

Automated Building Commissioning Analysis Tool (ABCAT)

This tool collects and compares whole-building energy use to modeled 'optimal' energy use, and identifies anomalies that point towards operational problems.

Diagnostic Agent for Building Operation (DABO^{TM})

This tool is an add-on module that uses the Building Automation System (BAS) to monitor performance at the sub-system level. It is designed as an aid for Ongoing Commissioning.

The basic architecture and functionality of each of these systems is described, along with case study examples of projects where they identified system problems and contributed to their solutions.

Conclusions

Commissioning Cost-Benefit

The data collected through this research project begins to characterize the various types of commissioning processes that are occurring in Annex member countries internationally. While data was often difficult to obtain, we expanded our knowledge in two key areas:

- The scope of the Cx process employed for new and existing buildings, and

- Characterization of issues discovered through the Cx process including system type, likely origin of issue (design, construction/installation, Operations and Maintenance (O&M), or capital improvement), issue type, and measures implemented.

While the project results begin to develop a qualitative picture for how commissioning is evolving internationally, quantitative results were less apparent. For example, data on commissioning costs and energy savings were highly variable. Falling short of the data collection goals set by Annex member country representatives, it was not possible to make strong conclusions about the cost-effectiveness of Cx internationally. However, progress was made towards understanding and categorizing the state of the commissioning industry for new and existing buildings in Annex member countries. While all countries have Cx research occurring, the majority of countries are in an early adopter phase of industry development. Only a few countries can be categorized as having a developing commissioning industry in which services are becoming more commonly obtained by owners.

New Construction Cx costs ranged from 0.06 US$/ft^2 to 2.57 US$/ft^2, (0.64US$/m^2 to 27.63 US$/m^2) suggesting the Cx process varied significantly and/or the way costs were attributed varies. Savings values were either not reported or considered unreliable as reference values, and so payback values were not calculated for New Construction Cx.

For Existing Building Cx, calculating simple payback results in a small data set (19 samples) but this can serve to illustrate a typical range of values (see following chart).

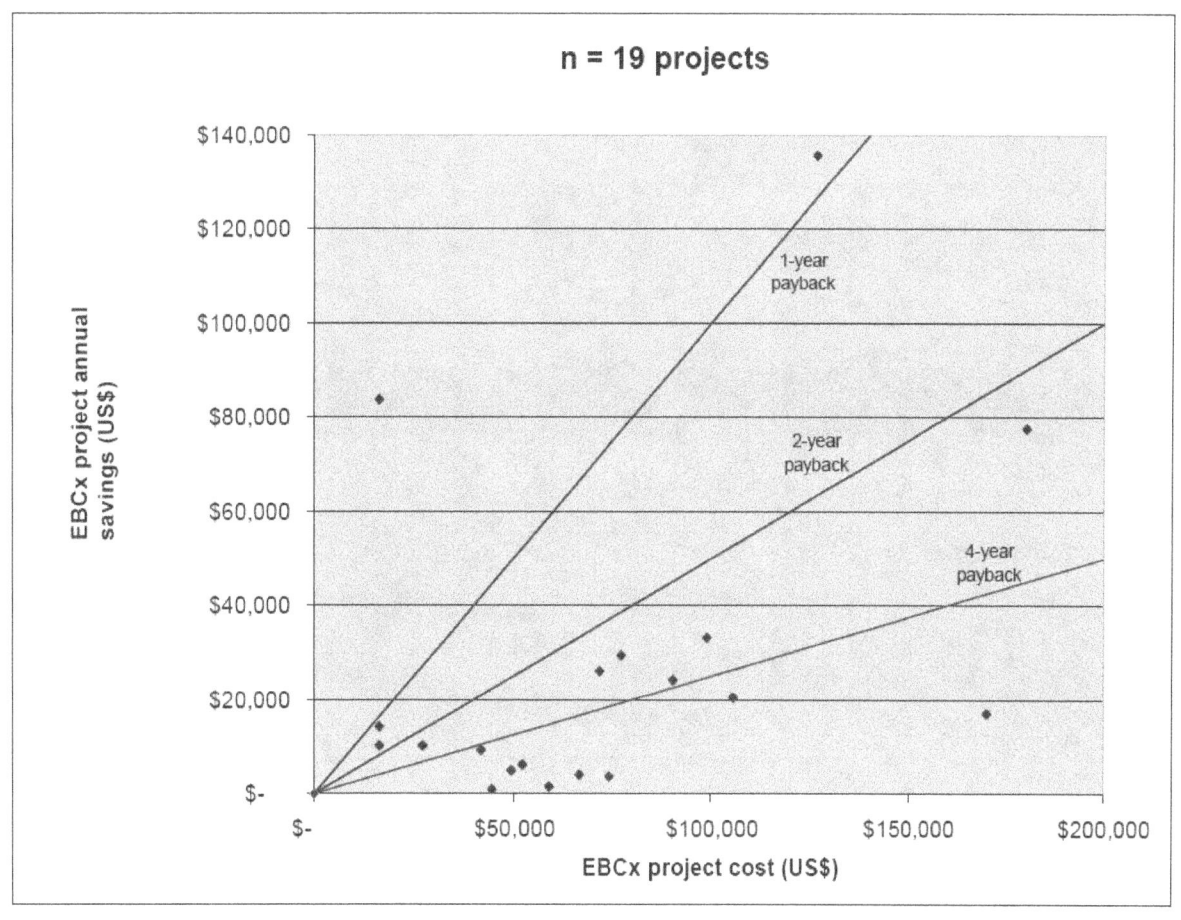

EBCx project costs vs. annual savings estimates

Project simple payback values ranged from 0.9 years to 45.7 years, with a median value of 3.7 years. Nine out of the 19 projects had a payback of greater than four years, and six had payback of between two and four years.

Higher payback is the result of *either* relatively high EBCx costs *or* relatively low resultant cost savings. For cases where payback was greater than four years in this study, the savings were relatively low (based on savings $/m^2 ($/ft^2), as opposed the costs being relatively high.

Persistence of Commissioning Benefits

Current information on persistence of commissioning energy savings in existing buildings may be summarized as:

- Savings persistence at the time of the study (3 years to 20 years after commissioning) ranged from about 50 % to 100 % in all but a handful of buildings.
- Average savings persistence at the time of the study was about 75 % of the original savings.
- The most dramatic savings degradation was caused by undetected mechanical or control component failures.
- Follow-up when needed has demonstrated persistence of commissioning savings for 7 years to 20 years in a small number of buildings. ABCAT and DABO™ have demonstrated the benefits of having tools to support persistence.
- Determination of savings using a "normalized annual consumption" as the basis for savings determination produced less variation in savings and persistence than found when the actual weather during the baseline and post-commissioning periods was used. It also suggested that use of calibrated simulation for baseline determination may provide more stable results.

1. INTRODUCTION

Building Commissioning[1] is a quality assurance process for the design, construction and operation of buildings. The formalized concept of commissioning originated in the UK in the 1960's and developed in Hong Kong and Canada in the 1980's. The process has been refined in individual countries and researchers are working with the building industry to merge the commissioning process with existing best practices to develop a more complete and streamlined approach to ensure that buildings operate as intended. Internationally, partnerships and collaborations to promote the commissioning process have accelerated awareness and adoption [Appendix B]. Although it is recognized as a valuable means to ensure that buildings reach their operating potential, the process is not widely adopted.

The principle barrier to market penetration is the perception of high cost of commissioning. Documenting the costs and the benefits, and disseminating that information is widely seen as critical to increasing the uptake of commissioning. The increased use of innovative, interacting, systems in low or zero energy buildings both increases the importance of commissioning and requires the development of commissioning methods and procedures for these systems. Furthermore, it is crucial that the benefits gained through commissioning persist over time so that building owners maximize their benefits.

The development of standardized methodologies for quantifying costs and benefits of commissioning and the evaluation of persistence of savings are seen as means to further increase market adoption of. This report presents the lessons learned in key studies, gathers and develops new information, and distills the information into a format that can useful in the development of a plan for future work.

This report summarizes the work completed between 2005 and 2009 through the International Energy Agency Annex 47 Subtask C: *Cost-Benefit and Persistence of Savings*, and is organized as follows:

Chapter 2: Commissioning Cost-Benefit
This chapter includes a literature review of existing cost-benefit methodologies, including recommendations for improvements.

Chapter 3: Commissioning Persistence
This chapter discusses methodologies for measuring persistence of commissioning benefits, summarizes example projects from a literature review, and reports results of persistence studies performed by Annex participants.

Chapter 4: Tools to Enhance Persistence
This chapter describes two tools that have been developed for the purpose of tracking persistence of building performance – ABCAT and DABO.

Chapter 5: International Commissioning Cost-Benefit and Persistence Report
This chapter summarizes the results of the Annex 47 cost-benefit data collection efforts with discussion of findings and recommendations for future work.

Chapter 6: Conclusions on Commissioning Cost-Benefit and Persistence

---

[1] Commissioning- Clarifying Owner's Project Requirements (OPR) from viewpoints of environment, energy and facility usage, and auditing and verifying different judgments, actions and documentations in the Commissioning Process (CxP) in order to realize a performance of building system requested in the OPR through the life of the building.
Initial Commissioning, Re-commissioning, Existing Building Commissioning, and Ongoing Commissioning are defined in the Glossary of Terms produced by ECBCS Annex 40.

2. COMMISSIONING COST-BENEFIT

2.1 Literature Review of Commissioning Cost-Benefit Methodologies and Data

2.1.1 Introduction

This chapter summarizes findings from a review of 12 commissioning cost-benefit studies. The chapter focuses specifically on the methodologies used to determine the costs and benefits of commissioning (See Appendix A for report on methodology). In order to maintain this focus, only studies that make their methodologies explicit have been included. The majority of methodologies that were analyzed are research studies of multiple buildings, and only a few are case studies of just one or two buildings. A more exhaustive list of studies that include cost and benefit data, but not an extensive methodological discussion, can be found in the bibliography.

These 12 studies represent a variety of formats and intentions, which were each created to meet the funder's goal. Among them are research reports, databases of cost-benefit information and a glossy, marketing-style brochure. Most of the research reports were undertaken to produce data to support utility and research programs and to help owners and commissioning providers gather the financial justification needed to implement New Building Commissioning (Cx) or Existing Building Commissioning (EBCx).

There is a significant difference in methodological framework between studies implemented as "one-time" or "snapshot" analyses, and those set up to continually collect and incorporate new data. It is probably true that any methodology can be implemented on a continuous basis if its funding is also continuous. However, data collection methodologies that facilitate data entry by allowing respondents to easily enter their own data and use an automated or semi-automated analysis tool are better positioned for ongoing analysis.

Table 2.1 summarizes the key cost and benefits as derived from the 12 studies. This indicates the minimum and maximum *average values* stated in the studies, rather than the complete range of values for all individual *projects* (which would have a larger range of values).

Table 2.1 Summary of average costs and benefits from 12 studies[i]

	New construction US$/m^2 (US$/ft^2)	Existing buildings US$/m^2 (US$/ft^2)
Cx / EBCx cost	$2.05 to $10.76 ($0.19 to $1.00)	$0.86 to $4.31 ($0.08 to $0.40)
Energy benefits	$0.54 to $6.89 ($0.05 to $0.64)	$1.18 to $2.80 ($0.11 to $0.26)
Non-energy benefits	$1.40 to $22.60 ($0.13 to $2.10)	$1.18 to $1.94 ($0.11 to $0.18)
Simple payback	4.8 years to 6.5 years	0.7 years to 3.2 years

[i] Costs and benefits are presented as ranges to demonstrate the variances in the studies examined. Median or average values are not presented because underlying methodologies differ widely and such figures would not reflect actual costs and benefits experienced by building owners.

Table 2.2 describes the studies, including their format, expected use and audience, and any caveats and considerations that might affect how their conclusions are interpreted. Four of the studies were originally conducted as research projects, funded by government agencies and a non-profit corporation.[2] Ten reported their findings in published conference papers[3] of the two that were never published as conference papers, one is a glossy brochure produced for marketing purposes and the other is an article written for subscribers to an energy research and information service.[4]

Table 2.2 also provides a side-by-side comparison of the studies' data and findings. They represent a wide range of methodological approaches and resulting data on the costs and benefits of commissioning. Their data ranges from case studies of one to six buildings to more extensive analyses of 16 to 21 buildings to two meta-analyses of data collected and analyzed by others, one of 44 buildings and the other of 175 buildings. Among building projects studied there is wide range in building size and type and in findings.

---

[2] Funders were: U.S. Department of Energy, U.S. Environmental Protection Agency, Lawrence Berkeley National Laboratory, Northwest Energy Efficiency Alliance.

[3] Conferences were: American Council for an Energy Efficient Economy Summer Study, National Conference on Building Commissioning.

[4] Energy information service is ESource.

Table 2.2. Description of cost-benefit studies (continued on next two pages)

Study/author	Format	Use/audience	Caveats and considerations
Stum, ECM Cx (1994)	Conference paper	Research	▪ Only energy conservation measure (ECM) commissioning is studied – not whole building commissioning. ▪ Utility program costs are included as a cost of commissioning. ▪ This early study does not address non-energy benefits (NEBs).
Piette, Energy Edge Cx (1995)	Technical report	This was one of the first studies to show savings concretely, and audience is program planners, technology developers.	▪ The study is focused on commissioning of ECMs in new construction, although additional unrelated deficiencies were reported. ▪ Some of the data collection and analysis were associated with a broader evaluation project.
Haasl, 5 Building Study (1996)	Conference Paper	Funded by the U.S. Environmental Protection Agency, Global Change Division, and the U.S. Department of Energy to help formulate energy conservation policy and programs.	▪ The study was an "Operations and Maintenance (O&M) investigative case study." ▪ No detail provided on the standard energy calculations or modeling scope. ▪ NEB analysis was still in progress. ▪ Stated objective was to "demonstrate that energy saving opportunities exist... and can be realized through improvements in O&M."
PECI/DOE, Deficiency Database (1996)	Research report	DOE-funded to document deficiencies found through Cx and EBCx.	▪ While technically not a cost-benefit methodology itself, this method offers insight into the value of incorporating a detailed deficiency database into any cost-benefit methodology. ▪ Savings data only available for 35 deficiencies. ▪ A deficiency database may be a lower cost version of a cost-benefit methodology. Typically Cx reports have some detail on measures at a findings level through an issues log or punchlists. Cx reports do not always have comparable detail regarding quantification of energy savings or non-energy savings and a cost accounting procedure. A deficiency database leads to an understanding of where the most common problems lie.
Gregerson, EBCx (1997)	Report for members of ESource	Audience was ESource members (utilities, ESCOs, Cx providers, researchers), to quantify a new field of efficiency opportunity	▪ Few reports cited measure costs and savings. Savings may be estimated, or as with the Texas LoanSTAR program (75 % of square footage in the study sample) per-building costs were estimated. ▪ The first major summary report on EBCx.
PECI, Brochure (1997)	Glossy brochure	Audience was owners and Cx providers, for marketing.	▪ Summary metrics by sector, Cx and EBCx mixed. ▪ Original data not available.
Altweis (2001)	Conference paper documenting methodology and detailed assumptions	Paper's audience was Cx providers, to encourage them to collect and report such data. Audience for data is owners and prospective customers.	▪ Very small sample size, suitable for case studies or research projects. ▪ A wide range in savings reported, due to highly varying assumptions (scenarios). Savings calculation methodology will vary from Cx Agent to Cx Agent, no standard calculation provided (although the methodology is conceptually well defined). ▪ No discussion of costs.

Study/author	Format	Use/audience	Caveats and considerations
Mills, Meta-analysis (2004)	Excel database, Research report	Statistical analysis for U.S. Dept. of Energy	▪ Largest Cx cost-benefit study to date. Focused on obtaining large number of projects to get a high-level view of Cx metrics. ▪ Relied on availability, quality, and comparability of different primary data sources. ▪ Majority of building information comes from a few sources, especially for EBCx. ▪ Merits of Cx should be assessed based on the cost-effectiveness of the proposed measures, not necessarily only on what was implemented. ▪ May inappropriately attribute or not attribute costs to the Cx process since cost accounting conventions are not always followed. ▪ May underestimate benefits because energy savings from all measures are not captured in Cx reports, NEBs are not usually expressed in monetary terms, and financial benefits in terms of increased net operating income (NOI) are rarely determined. Furthermore, in a few projects studied, measured savings exceeded predicted savings. ▪ Time consuming to gather project information from secondary sources and interpret it, as opposed to having the cost-benefit data entered by the people involved with the project.
SBW, Northwest Cx & EBCx (2004)	Research report	Utility program evaluation	▪ Cost calculations include many costs associated with Cx, so figures may be higher than other studies. ▪ NEBs calculations based on opinion of team members (willingness to pay and/or perceived value).
California Commissioning Collaborative's Cxdatabase.com (2004)	SQL database, exportable to Excel One-page "datasheet" on each project Conference paper describing database	Researchers – data that supports Cx research and utility incentive programs Owners – defining the value of Cx to their business through data and case studies Providers – third party source from which to give owners information. Help raise the bar for Cx documentation of results	▪ Data was stored as-entered by respondents – no analysis performed unless brought in by outside researchers ▪ Datasheet was a one-page summary form automatically populated by data entered by respondents. ▪ Little population as of February, 2006 – database is no longer online, and is not actively updated. ▪ While the original vision for this data included creation of case studies, none were created.

Study/Author	Format	Use/Audience	Caveats and considerations
Moore et al. California EBCx programs lessons learned (2008)	Presentation paper for ACEEE Summer Study (pdf)	Energy efficiency program implementers	<ul><li>Report-out on 4 California EBCx Programs, covering 60 million square feet of commercial buildings (approximately 200 buildings)</li><li>Describes program process experiences as well as presenting quantitative data</li><li>Reports on:<ul><li>kWh savings recommended after investigation, compared with savings from measures selected by owner for investigation</li><li>Simple payback for uncovered measures and measures selected by owner for implementation</li><li>Breakdown of savings by measure type</li><li>Simple payback by measure type</li><li>Savings by building type</li><li>Relationship between building EUI and EBCx savings</li></ul></li><li>All programs were managed by PECI under the same basic program design, so scope of work and approach to energy savings calculations are consistent across all programs. Commissioning provider costs and driven by the utility program design/budget, and so are consistent across the dataset.</li><li>Much of the data is based on outcomes of EBCx investigations; at the time the paper was created, a high proportion of the projects had not completed implementation</li></ul>
PECI & Summit Building Engineering 2007 California EBCx Market Characterization (2008)	Report available from California Commissioning Collaborative www.cacx.org	Program implementers, providers	<ul><li>Estimates of potential costs and savings for EBCx in California, along with an assessment of how many EBCx providers would be required to fulfill that potential</li><li>Projects used for the study were all under California utility programs, and so savings calculation approaches are quite consistent across the whole data set</li><li>Forecasts based on a very large number of ongoing projects under utility programs; only a small number of these projects were complete at the time of the report.</li></ul>

2.1.2. Data Collection Strategies

By far the most common data collection method is the Researcher-Driven Model, in which a researcher was tasked with collecting and analyzing data. In more than 90 % of the studies a researcher was wholly or partially responsible for collecting documentation and data produced by others. In the handful that differed from this model, data collection was usually done "in house," because the researcher also served as the commissioning provider on the projects that were studied. In two cases, however, data collection was accomplished through use of a database allowing providers and owners to submit data independent of the researcher. In two cases, data came from databases created for utility existing building commissioning programs, where project data was entered by program administrators as a part of the normal program process. A comparison of the data collection methodologies used in 13 different studies reviewed here is presented in Table 2.3. The discussion of the three cost methodology types (simple, moderate, and complex) shown in Table 2.3 is presented in Sec. 2.1.3.

The Researcher-Driven Model

In 10 of the 13 studies, the data collection strategy was driven by a researcher who collected commissioning project information and produced a cost-benefit report. In eight of those ten, the researchers relied heavily on project documentation, primarily the commissioning provider's *Final Report*. Other documentation consulted included construction documents (for new buildings), issue logs and change orders. In more than half of the studies, other types of information were used to supplement written documentation. They include telephone surveys with key team members (two studies) and onsite inspection and monitoring (three studies).

When telephone surveys or interviews were employed, they were often used to gather data on non-energy benefits (NEBs). This is logical, given that NEBs are hardest to measure using commissioning or building documentation because they depend most on the experiences of the people who manage and occupy the building. In fact, there are two studies in which researchers were only interested in NEBs and in which the only source of data were telephone surveys and detailed interviews – no project documentation was collected (Haasl 1996; Bicknell 2004).

Among the ten studies that employed the researcher-driven model, there is much variation in the amount of data studied and level of detail collected, the logistics of obtaining documentation, and in the supplemental types of data and the data collection strategies.

- **Quantity of data** varies from case studies of a single project to mid-range studies of five, six, 21 and 44 projects to two large studies of 175 projects each.
- **Level of detail** ranges from whole-project level metrics to metrics for individual issues.
- **Logistics of obtaining documentation** includes submission by a utility that collected all the documentation and turned it over to the researcher, submission by owners and providers directly to the researcher, and the gathering of documentation by the researchers from commissioning providers and other researchers.
- **Supplemental data** includes telephone surveys and onsite inspections.

In the researcher-driven model, data collection almost always takes place after the commissioning projects are complete and documentation finished. As a result, the effort required and the data quality depends almost entirely on the diligence of the parties responsible for producing the documentation (usually the commissioning provider, general contractor or testing and balancing agent). Time is also an issue. The closer the study is to project close-out, the more likely it is that project documentation will be available and in good condition, and that the important parties will be able to answer any questions.

The Provider-Driven Model

In two studies, the researcher and commissioning provider were one and the same. As a result, their studies were able to utilize very detailed data collected throughout the commissioning project. However, only a few projects were included, leading to these studies' designation as case studies rather than statistically significant research studies (Haasl 1996; Altweis 2001).

The Research Database Model

In two studies, researchers created interactive databases to collect commissioning project data. In one, the database was created through a collaborative effort in which multiple researchers and commissioning providers helped define required and minimum inputs. The database itself was created as an online application, meaning it was accessible on the Internet. Thus once it was released, commissioning providers could use it to enter information about their projects in real-time[5]. In the other, a database of categories deficiencies was developed (PECI/DOE 1996).

A significant advantage to the database model of data collection is the ability of the researcher's needs to influence the commissioning provider's data retention efforts. Because providers know up-front what data the researcher wants, it can be supplied immediately while the documents are still available and the project is fresh in the provider's mind. On the negative side, a database alone is incapable of performing analysis, and this model requires funding for several things: design and programming of the database, a researcher to analyze the data or work with programmers to build analysis functionality into the database, marketing of the database to the provider audience, ongoing database maintenance and support, and perhaps even funding to compensate providers for entering data.

The Utility Program Database Model

In two cases, data came from databases created for utility EBCx programs, where project data was entered by program administrators as a part of the normal program process. Data entry came from two sources: 1) Data entered directly into database by program administrators (including EBCx provider costs) and 2) Data uploaded from each project's Master list of Findings, which lists descriptions, savings, and costs for each individual measure. EBCx providers enter data into the Master list of Findings spreadsheet, and program administrators upload this spreadsheet to the program database. Each project's Master List of Findings, undergoes program review (along with all supporting documentation, collected data and energy savings calculations).

The main advantages of this data are:
- Cost and savings information is available for each individual measure.
- All data underwent a rigorous quality control process that maximized consistency across the dataset
- Data is available for a high quantity of projects within a single U.S. state.

The main disadvantages of this data are:
- There is no non-energy benefits data included in the utility program databases.
- Data available at the time the two reference reports were created was predominantly based on unfinished projects, so it does not include much data on installed measures (only measures *selected* for implementation by building owners). Data on installed measures for the California projects should be available by the end of 2009.
- EBCx provider costs are driven by utility program designs/budgets. This is not necessarily a *disadvantage*, but it does artificially influence the overall project costs.

Explanation of Estimated Effort

Table 2.4 includes a column for "Estimated effort need to obtain and enter data." The amount of time and difficulty required to both collect and submit project data is estimated as either **low, medium, high, or variable**. These rankings are not independently defined. Rather, they reflect the authors' estimate of the relative effort required to gather data according to the study's methodology, as compared to the other studies in this report. Thus a study with a "high" effort ranking was judged to employ a more time- and effort-intensive collection methodology than those deemed "low" or "medium."

---

[5] This database is no longer online, and is not being maintained.

Table 2.3. Cost and benefit data (continued on next 2 pages)[6]

Study/Year	# of buildings	Total and median bldg. size	Costs (US$)		Energy benefits		Non-energy benefits (NEBs)		Cost effectiveness	
			Cx	EBCx	Cx	EBCx	Cx	EBCx	Cx	EBCx
Stum, ECM Cx (1994)	6	218 722 ft² 20 320 m²		$3 060 overall $0.041/ ECM $0.08/ft² $0.86/m² **Simple**		37 412 kWh/y 5.3 % of orig. ECM Unrealized: 7.9 % **Moderate**		N/A		Recovered savings: $0.033/kWh Recovered + unrealized savings: $0.02/kWh
Piette, Energy Edge Cx (1995)	16	849 800 ft² 78 949 m² 27 000 ft² (median) 2 508 m² (median)	$0.19/ft² $2.05/m² **Simple**	N/A	9.48 kWh/ft²·y 102.04 kWh/m² $0.64/ft²·y $6.89/m²y **Complex**	N/A	not quantified	N/A	Simple payback: average 13.7 y; median: 6.5 y	N/A
Haasl, 5 Building Study (1996)	5	837 000 ft² 77 760 m² 1 313 197 ft² (median) 122 000 m² (median)	N/A	$0.11/ft² $1.18 m² **Simple**	N/A	$0.11/ft² $1.18 m² **Simple**	N/A	Not quantified	N/A	Simple payback: 10 months
PECI/DOE, Deficiency Database (1996)	16 Cx 28 EBCx (44 total)	4 million ft² 367 896 m² 67 000 ft² 6 224 m² (median)	N/A	N/A	83 % of all deficiencies related to energy	92 % of operational deficiencies impact energy. Avg savings/ deficiency = $892/y **Moderate**	51 % of all deficiencies related to reliability and maintenance	25 % of deficiencies related to comfort	N/A	N/A

---

[6] Dollar amounts have not been normalized to a common year. Methodological complexity listed in **bold**.

Study/Year	# of buildings	Total and median bldg. size	Costs (US$)		Energy benefits		Non-energy benefits (NEBs)		Cost effectiveness	
			Cx	EBCx	Cx	EBCx	Cx	EBCx	Cx	EBCx
Gregerson, EBCx (1997)	44	9 million ft² 836 127 m²	N/A	Approx $20 000 $0.19/ft² $2.05/m² **Simple**	N/A	Avg $98 000 Med: $41 000 19.2 % avg savings $0.49/ft² $5.27/m² **Moderate**	N/A	Not assessed	N/A	Simple payback: 0.9 years [7]
PECI, Brochure (1997)	75 Cx and EBCx not separated	Not available	Median: $0.15/ft² $1.61/m²		No standardized metric **Moderate**		Improved (% of buildings): Thermal comfort: 42 % System function: 44 % Indoor air quality: 23 % O&M: 42 % **Simple**		Not assessed	
Altweis (2001)	1	14 350 ft² 1 333 m²	Not reported	N/A	Up to $0.13/ft²·y $1.40/m²	N/A	$0.17 to $2.10/ft²·y $1.83 to $22.60/m²y **Moderate**	N/A	Not reported	N/A
Heinemeier, Schools Cx (2004)	1	N/A: methodology but no results.	N/A	N/A	N/A	N/A	N/A	N/A	N/A	N/A

[7] Energy intensive buildings and even most of the efficient buildings had paybacks of less than two years.

Study/ Year	# of buildings	Total and median bldg. size	Costs (US$)		Energy benefits		Non-energy benefits (NEBs)		Cost effectiveness	
			Cx	EBCx	Cx	EBCx	Cx	EBCx	Cx	EBCx
Mills, Meta-analysis (2004)	175 projects EBCx: 106 Cx: 69	30 400 000 ft² 2 824 252 m² total 69 500 ft² 6 457 m² (median Cx) 151 000 ft² 14 028 m² (median EBCx)	$74 267 $1.00/ft² $10.76/m² [0.6 % construc-tion cost] **Moderate**	$33 696 $0.27/ft² 2.90 m² **Moderate**	$2533/y $0.05/ft²·y $0.54/m²·y **Moderate**	$44 629/y $0.26/ft²·y $2.80m²·y **Moderate**	$51 000/y $1.24/ft²·y $13.35m²·y **Moderate**	$17 000/y $0.18/ft²·y $1.94 m²·y **Moderate**	Simple payback: 4.8 y[8]	Simple payback: 0.7 y[9]
SBW, Northwest Cx & EBCx (2004)	21	2.2 million ft²[10] 204 386 m²	$71 791[10] $0.85/ft² $9.15/m² **Complex**	$22 053 $0.31/ft² $3.34/m² **Complex**	$9 856/y $0.09/ft²·y $.97/m²·y **Complex**	$13 678/ year $0.14/ft²·y $1.51m²·y **Complex**	$13 609 (one-time) $0.13/ft² $1.40/m². **Complex**	$10 534 (one-time) $0.11/ft² $1.18/m² **Complex**	*Direct pay-back: 7.5 y* *Total simple payback: 6.1 y*	*Direct pay-back: 4.0 y* *Total simple payback: 3.2 y*
Moore et al. (2008)	21	Not stated, but 21 projects total approx. 9 million kWh annual savings forecast		Not reported		Average 6.9 % **Complex**	Not reported			*84 % of savings with <2-year payback (owner's implementa-tion costs only)*

[8] Standard energy prices corrected for inflation. Data normalized to $2003.
[9] Standard energy prices corrected for inflation Data normalized to $2003.
[10] Standard energy prices corrected for inflation Data normalized to $2003.
Costs for this study include only Cx provider fees – although payback information includes additional costs, for example, costs to other parties.

Study/ Year	# of buildings	Total and median bldg. size	Costs			Energy benefits			Non-energy benefits (NEBs)			Cost effectiveness		
---	---	---	Cx	EBCx		Cx	EBCx		Cx	EBCx		Cx	EBCx
PECI & Summit Building Engineering (2008)		Market characterization study forecasts 70 million ft^2 annual potential for EBCx in California (6.5 million). m^2		$0.39/ft^2 4.20/m^2 **Complex**			0.96 kWh/ft^2·y 10.3 kWh/m^2·y **Complex**			Not reported			*Mean 3-year payback forecast*
California Commissioning Collaborative's Cxdatabase.com (2004)	Two surveys were completed and five were in progress. Not assessed at this time due to lack of data and funding.												

Table 2.4 Comparison of data collection methodology (continued on next page)

Study/Author	Level of detail	Data sources	Collection process	Timing of data collection	Data storage	Estimated effort needed to obtain and enter data
Stum, ECM Cx (1994)	Only looked at ECMs meriting greater resources (i.e., VFDs, economizers)	Inspection reports, onsite inspection of ECMs in small comm. and retail	Reports provided by utility, onsite work done by authors.	Concurrent with and immediately following Cx activities	N/A	**Medium** Onsite inspections but only of a few measures
Piette, Energy Edge Cx (1995)	Very detailed data collection on building characteristics to develop models,	Commissioning report, on-site data collection.	Extensive evaluation project.	Within one or two years after construction.	Unix-based database.	**High** Data for simulation.
Haasl, 5 Building Study (1996)	Data required for standard calculations and simulations.	Provider collected.	Data collected through EBCx process, including two weeks of monitored data on key systems.	Collected during EBCx process.	Not described.	**Medium** Building and system characteristic data needed for modeling and calculation, and monitored data.
PECI/DOE, Deficiency Database (1996)	Findings level detail.	Half of data directly entered by Cx provider. Half by researcher (paid).	Review of final commissioning reports and issues logs.	Retrospective.	Database (Excel)	**Variable** Depends on availability and organization of necessary info in Cx documentation.
Gregerson, EBCx (1997)	No detail other than metrics on a project-level. No measure-level detail.	Four Cx providers. 70 % from TAMU and 25 % from PECI	Building characteristics, EUI, and cost and energy savings figures requested from Cx provider by researcher.	Retrospective from final Cx reports.	Not specified.	**Low** Very minimal data collected (although retroactive so it may be difficult to obtain.)
PECI, Brochure (1997)	High level metrics	Cx provider	Phone interviews	Retrospective	Not specified	**Low** Minimal data collected (although retroactive so it may be difficult to obtain)

Study/Author	Level of detail	Data sources	Collection process	Timing of data collection	Data storage	Estimated effort needed to obtain and enter data
Altweis (2001)	Case study level data. Extensive detailed information required from project. Not difficult for the commissioning agent to obtain.	Commissioning agent collected information and conducted benefits analysis.	Commissioning provider review of notes and project documents.	Throughout the project.	Not specified.	**Low**, for commissioning provider. **High** for anyone else.
Heinemeier, Schools Cx (2004)	Report showed a great deal of detail, but the intent is to define metrics that are easily gathered, from review of construction documents.	Complete construction documents.	Researcher obtained a copy of and reviewed all construction documents.	After the project was complete.	Not specified.	**Medium** Somewhat time-consuming to review all documents.
Mills, Meta-analysis (2004)	Based on the documentation available. Where little available, at minimum, project-level info was entered.	A few Cx providers and researchers entered many projects (paid). Smaller number of projects from unpaid Cx providers.	Review of past studies and final Cx reports/issues logs.	Projects completed between 1993 and 2004 Retrospective from final Cx reports and previous studies.	Excel spreadsheet	**Variable** Depends on availability and organization of necessary info in Cx documentation.
SBW, Northwest Cx & EBCx (2004)	Identified all issues/findings, selected only significant and resolved issues	Extensive project documentation and surveys (both unpaid).	Project materials submitted by owner and telephone surveys with team.	While projects underway and within 1 year after close-out (early 2003)	Database (no specific software identified)	**High**
Moore et al. (2008)	High level utility program data reported.	Utility program databases	Paper created by program administrator with direct access to all programs' data	Data compiled Summer 2008	Excel spreadsheet downloaded from program databases	**Medium** Data is readily available (but need to filter data for non-standard projects)
PECI & Summit Building Engineering (2008)	High level forecasts of market potential	Data collected from various California utility program implementers	Data requests to all implementers. Data then adjusted for easier cross-comparison	Early 2008	Spreadsheet	**High**

Study/Author	Level of detail	Data sources	Collection process	Timing of data collection	Data storage	Estimated effort needed to obtain and enter data
California Commissioning Collaborative's Cxdatabase.com (2004)	Three findings required, can accommodate unlimited number	Cx provider or owner. (unpaid, but funding for entering data desired, requirement was written into scope of some projects)	Respondent gathers data and enters into online forms.	Intended to be completed during project or immediately after completion. Can be completed at any time, if data is available.	Custom-built online database – project took several months at a cost of approx. $20,000. (www.Cxdatabase.org)	**Variable** Depends on whether respondent was aware of data requirements during the project and the quality of documentation.

2.1.3 Methodologies for Determining Costs, Energy Benefits and Non-Energy Benefits

Commissioning Costs

There is no widely used methodology for determining commissioning costs. To assist in the evaluation process, this report distinguishes three levels of complexity in cost methodologies: simple, moderate and complex.

Table 2.5 summarizes their differences in terms of which costs are included and if the costs are validated.

Table 2.5 Comparison of cost methodologies

	Cx provider's fee	Resolution costs (EBCx)	Costs to other parties	Validation method
Simple	X	X		
Moderate	X	X	X	
Complex	X	X	X	X

Of the 12 studies examined, all include the commissioning provider's fee as a cost of commissioning. Some include additional costs, for example, costs to other parties, although each study defines these costs differently. Only one study, with a complex methodology, makes an attempt to validate cost data by checking the respondent's data for consistency and to make sure cost figures fell within what researchers defined as a "reasonable range." (SBW Consulting 2004)

In general, the average cost of commissioning per square foot *increases* as the study's cost methodology *increases in complexity*. As Table 2.6 shows, in existing buildings the cost of commissioning steps upward as the methodology becomes more complex. In the case of new buildings, the cost of commissioning has a generally higher trend; however, complex methodologies returned an average cost per square foot (^{m2}) slightly lower than moderate methodologies. Although not conclusive, it seems likely that the reported cost increases because complex methodologies account for costs incurred by several parties, whereas simple methodologies usually only account for the commissioning provider's fee.

Table 2.6 Average cost of commissioning by methodological complexity

Methodology	Existing buildings		New buildings	
	# of buildings	Average cost (US$)	# of buildings	Average cost(US$)
Complex	8	$0.31/ft^2 $3.34/m^2	13	$0.85/ft^2 $9.15/m^2
Moderate	106	$0.27/ft^2 $2.91/m^2	69	$1.00/ft^2 $10.76/m^2
Simple	50	$0.18/ft^2 $1.90/m^2	16	$0.19/ft^2 $2.05/m^2

A more detailed discussion of cost methodology types, with examples, follows.

Simple. A simple methodology uses only one or two cost categories to arrive at the overall cost of commissioning. Usually these cost figures are relatively easy to obtain. Examples include the commissioning provider's fee and the cost to resolve an issue. In simple methodologies, these cost figures are self-reported and the study makes little or no attempt to validate the data.

An example of a simple methodology can be found in two early studies: Piette (1995) and Stum (1994). Piette calculated the cost of commissioning by taking a percentage of the overall energy efficiency measure cost. Stum defined the cost of commissioning as the self-reported commissioning provider's fee plus the administrative costs of the utility commissioning program that funded the projects.

Moderate. A moderately complex methodology uses more than two cost categories to arrive at the overall cost of commissioning. For example, cost categories could include incremental costs to all parties, travel costs, and negative impacts like increased change orders. Moderately complex methodologies include a broader array of costs in the cost of commissioning than a simple methodology, although the study stops short of applying a validation process to the data. The methodology may include differences in cost accounting between Cx and EBCx.

An example of a moderate methodology can be found in Mills et al. (2004) and in Cxdatabase.com (2004). Mills et al.'s cost definition includes several figures: the provider fee, the coordination costs incurred by other parties and on EBCx projects, the resolution costs. Cxdatabase.com's cost definition includes the provider fee, incremental costs incurred by other parties, the cost of O&M staff participation (if specified by the owner) and on EBCx projects, the resolution costs. Neither Mills et al. nor Cxdatabase.com makes any attempt to verify the cost figures reported in project documentation or by respondents.

Complex. Like a moderate methodology, a complex methodology differentiates several categories of commissioning costs. However, studies employing complex methodologies do attempt to validate cost figures.

An example of a complex methodology is found in SBW Consulting (2004). Cost includes the provider fee, incremental fee increases for other parties, travel expenses and resolution costs. Researchers used a telephone survey to ask key commissioning team members 1) if they increased their bid for the project to account for commissioning activities and 2) if there were any significant non-labor costs associated with commissioning. Respondents were then asked to attach a dollar amount to each. If the respondent was unable to provide a dollar figure, researchers asked them to estimate the additional labor hours and provide a labor rate, from which researchers calculated the incremental cost to that respondent. As a quality assurance measure, researchers also evaluated whether the data supplied by respondents "were consistent and fell within reasonable ranges" (SBW Consulting 2004).

Energy Savings

There are a variety of methods for determining energy savings from commissioning. This study evaluates energy benefit methodologies, like cost methodologies, according to their level of complexity.

Table 2.7 Comparison of energy savings methodologies

	Issues ID and/or baseline comparison[11]	Energy calcula-tions	Energy modeling	Normaliza-tion of energy data	Attention to measure interactions	Validation method
Simple	X					
Moderate	X	/	/	/		/
Complex	X	X	X	X	X	X

Key: X = always present; / = sometimes, but not always present

Of the 12 studies examined, all used either an issues identification or baseline comparison method to determine energy savings. Moderate methodologies employed some form of energy calculations, modeling, or data normalization. Only complex methodologies were attentive to measure interactions and data validation.

The following is a more detailed discussion of energy benefit methodology types, with examples.

Simple. In existing buildings, simple methodologies compare before and after energy consumption without normalization of data. They may also obtain information directly from the building owner or manager regarding energy savings or comparisons of performance.

Examples of simple energy benefit methodologies are found in Cxdatabase.com (2004) and Heinemeier (2004). Cxdatabase.com asked survey respondents to provide energy savings numbers for each reported finding. Respondents were asked to also provide the calculations they used to arrive at the figures, but this information was not required. No standardized process for calculating energy savings was created. In Heinemeier's methodology, energy use per square foot of commissioned buildings was compared to those that were not commissioned. Building pairs were of similar size and type, and monthly utility bills were used to gather energy use data. Commissioned building energy use was also compared against standardized benchmarks.

Moderate. Moderately complex methodologies use project documentation to identify significant commissioning findings/issues that have been resolved, and then use engineering calculations or parametric modeling to determine the energy benefit. A validation process using measured data may be, but is not necessarily, applied. Moderately complex methodologies may also apply normalization techniques to before and after energy consumption.

An example of a moderately complex energy benefit methodology is found in the SBW Consulting (2004). Researchers used a three-step process, shown in Fig. 2.1, to identify issues that resulted in a "stream" of energy and/or non-energy benefits. First they used project documentation to identify all issues. Then they determined which issues were "significant" relative to their affect on total building area or occupants, resolution cost and/or long term impact. (SBW Consulting 2004) Of significant issues, they determined through documentation and/or telephone surveys which issues had been or would likely be resolved. Energy and non-energy benefits were only calculated for issues deemed significant and resolved.

---

[11] Two methods for determining energy (and non-energy) savings are *issue identification* and *baseline comparison*. In *issue identification* energy savings are determined first at the issue level and then added to arrive at the total savings for the project. The *baseline comparison* method looks only at whole-building energy benefits. The researcher establishes the building's "baseline" energy use and then compares it to energy use after commissioning. This method can be a more straightforward process in existing building projects, where there is a "before" snapshot. In new construction it is more difficult because the "baseline" is hypothetical and must be simulated.

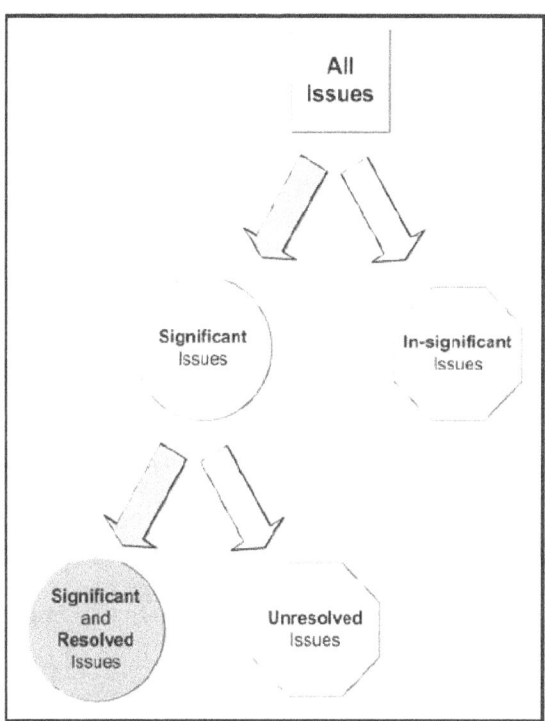

Figure 2.1 Sample issue identification methodology (SBW Consulting, 2004)

Complex. Studies utilizing complex methodologies employ detailed engineering calculations or models to estimate energy savings. Examples range from detailed building simulations that require extensive information about building characteristics to very detailed engineering calculations based on measured data. Complex methodologies for new construction commissioning benefits address nuances such as the range of assumptions that go into the hypothetical baseline (i.e., what is assumed to have occurred without commissioning). More complex methods also address the interaction between commissioning measures, and the interaction with related activities like energy retrofits. Results can be reported per measure, or for a whole building (which is not simply the sum of individual measures).

An example of a complex energy benefit methodology is found in Altweis (2001). This study used engineering calculations to estimate energy use both with and without identified findings/issues. Researchers developed both a "most likely" and a "least cost solution" scenario, depending on assumptions about what would have occurred in the absence of commissioning.

Non-Energy Benefits (NEBs)

In the assessment of non-energy benefits there is a great diversity of methodologies in use. Here again, in Table 2.8, non-energy benefit methodologies are classified according to their complexity.

Table 2.8 Comparison of non-energy benefit methodologies

	Monetary value NOT assigned	Monetary value assigned	Monetary value validated
Simple	X		
Moderate		X	
Complex		X	X

Of the 12 studies examined, 5 do not assign a monetary value to NEBs. Those that do use methodologies ranging from simple processes that do not employ standard calculations or checks on respondents' information to a highly complex system in which NEB dollar values are calculated several different ways and the most conservative number selected. Here too, only the most complex methodologies attempt to validate the data.

Contrary to the direct relationship between methodological complexity and commissioning costs, with NEBs there appears to be an inverse relationship: the more complex the methodology the *lower* the monetary benefit reported. This holds true for both new and existing building commissioning.

It is apparent that methodology significantly impacts reported non-energy benefits. The study employing avoided cost calculations (a moderate methodology) returned higher savings than the study that determined the owner's perceived value of the benefit (a complex methodology). This is due to a fundamental difference between these two methodologies. Although both methods are hypothetical, avoided cost calculations estimate the full cost that *would have been incurred*, had the benefit not been received. Whereas the owner's perceived value is the amount the owner is *willing to pay* for the benefit – often less than the avoided cost. Further study in this area is clearly needed to determine how the non-energy benefit valuation method relates to the goals of the cost-benefit methodology.

Table 2.9 NEBs of commissioning by methodological complexity [i]

Methodology	New Buildings		Existing Buildings	
	Number of Buildings	Average NEB Savings (US$)	Number of Buildings	Average NEB Savings (US$)
Complex	13	$0.13/\text{ft}^2$ $1.40/\text{m}^2$	8	$0.11/\text{ft}^2$ $1.18/\text{m}^2$
Moderate	23	$0.17/\text{ft}^2$ - $6.96/\text{ft}^2$ $1.83/\text{m}^2$ - $74.92/\text{m}^2$	10	$0.10/\text{ft}^2$ - $0.45/\text{ft}^2$ $1.08/\text{m}^2$ - $4.84/\text{m}^2$
Simple	no data		no data	

[i] Moderate data is presented as a range because a validation method was not employed.

The following is a more detailed discussion of non-energy benefit methodology types, with examples.

Simple. An example of a simple methodology for assessing non-energy benefits is found in Cxdatabase.com (2004). Here, respondents are asked to identify which benefits they received and have the option, but not the requirement, to supply a dollar value for the benefit. No standardized calculations are employed, and there is no process for evaluating the dollar values supplied by respondents.

Moderate. An example of a moderately complex methodology for calculating NEBs is found in Mills et al. (2004). Here, the researcher arrived at the NEB dollar value by adding the first-cost dollar value of non-energy savings and the ongoing labor cost savings, estimated as labor hours saved. Other NEBs were accounted for using a Yes/No checklist with an estimated dollar value supplied optionally.

Complex. An example of a complex methodology for assessing NEBs is found in SBW Consulting (2004), see Fig. 2.2. Researchers developed three different ways to assign a dollar value to a "stream" of benefits flowing from a specific finding/issue, and then used the most conservative (lowest dollar value) estimate. All three calculations were based on the responses of key commissioning team members given in telephone surveys. (See Table 2.10 for additional details).

1. Willingness to pay (WTP)
Survey question:

"If all the non-energy benefits (and negative effects) that we talked about were taken away, what do you think would be the maximum amount you would be willing to pay to get back those benefits, on an annual or monthly basis?"

2. Sum of individual computed benefits
Respondents asked to compare the commissioning provider's fee to the specific commissioning benefit. All benefits are then summed for a total NEB value. An example survey question:

"Would you say that compared to your annualized commissioning costs, the contractor call-backs are...about 10 % more valuable, about 1 to 1.5 times more valuable, twice as valuable, more than twice as valuable?" Or, "Don't know/refused."

3. Overall net value
Respondents were asked to identify significant impacts from a given list (e.g., reducing operational deficiencies). Their responses were weighted, to give the opinions of providers and facility staff more importance than contractors and designers. The gross dollar value of each impact was then multiplied by the importance factor.

Figure 2.2 Sample complex NEB methodology (SBW Consulting, 2004)

Table 2.10 Overview of cost-benefit evaluation methodologies (continued on next 3 pages)

Study/Author	Costs		Energy Benefits		Non-energy Benefits (NEBs)		Persistence Assumptions	
Stum, ECM Cx (1994)	Simple	Cost of Cx provider services + program administration costs	Moderate	Engineering calculations and computer simulations from original ECM savings predictions	N/A		N/A	
Piette, Energy Edge Cx (1995)	Simple	Percent of overall energy efficiency measure cost, costs by energy-efficiency measure are used when available.	Complex	Used deficiency identification and post-construction utility data to tune as-built simulation models. Deficiencies were "removed" from the models to determine savings. Deficiencies categorized as Directly, Indirectly, and Un-related to the ECM, and Static or Dynamic.	Simple	Categorized as control, indoor environmental quality (IEQ), equipment life, O&M. No value assigned.	Simple (but the modeling itself was not simple)	"Low" and "High" lifetime scenarios defined in the modeling.
Haasl, 5 Building Study (1996)	N/A	N/A	Simple	Categorize deficiencies into deficiency type (maintenance, documentation, training, operations, installation, design); HVAC subsystem; and affected component. Additional categorization for controls related findings.	Simple	All deficiencies were categorized, including non-energy related deficiencies. In total (Cx and EBCx), 51 % of all deficiencies related to reliability and maintenance (25 % significantly related). In total (Cx and EBCx), 25 % of all deficiencies related to comfort (13 % significantly related).	N/A	

Study/Author	Costs		Energy Benefits		Non-energy Benefits (NEBs)		Persistence Assumptions	
PECI/DOE, Deficiency Database (1996)	Moderate	Categorized as Assessment or Implementa-tion. Included consultant, contractor, and building staff time, as well as parts and lease costs for monitoring equipment. Did not include "research" related costs.	Moderate	"Potential" energy savings reported. A combination of standard engineering calculations and DOE2 building simulations, with short-term diagnostic monitored data used to inform the calculations or model.	Moderate	Only includes extended equipment life, which is the most easily quantified effect. Categorized as extended equipment life through reduced hours of operation and through reduced short cycling. Calculated based on assumptions of reduced hours, reduced lifetime through short cycling, and nominal life.	N/A	N/A
Gregerson, EBCx (1997)	Simple	Costs reported or estimated by each Cx provider. Costs include Cx fee, monitoring costs, and the cost of implementing measures except for in-house facility staff time during normal working hours.	Moderate	Project documentation may have utilized engineering calculations, models, or pre- and post-consumption measurement to quantify savings. Report notes that rigor with which energy savings were calculated varies significantly.	N/A		N/A	
PECI, Brochure (1997)	Simple	Cost range and median cost.	Simple	Savings range by building type. Conducted phone interviews.	Moderate	Identified NEB quantitatively for many different categories	N/A	

Study/Author	Costs	Energy Benefits		Non-energy Benefits (NEBs)		Persistence Assumptions	
Altweis (2001)	N/A	Complex	Used engineering calculations to estimate energy used with and without identified deficiencies. Provided *Most Likely and Least Cost Solution* scenarios, depending on assumptions for what would have occurred absent commissioning.	Moderate	Used simple calculations and extensive assumptions to estimate impacts to factors such as lost productivity, lost sales due to late building completion and equipment replacement. Provided *Most Likely and Least Cost Solution* scenarios, depending on assumptions for what would have occurred absent commissioning.	Simple	Lifetime assumed by measure and benefit (most are first-year impacts or flat over assumed lifetime).
Heinemeier, Schools Cx (2004)	N/A.	Simple	Comparison of monthly utility bills (electricity and gas (kBtu) per square foot), between commissioned and uncommissioned buildings (well matched pair or large sample size recommended), also comparing commissioned buildings with benchmarks (e.g., CBECS).	Moderate	Comparison of well-defined metrics collected during construction and operation phases, between commissioned and uncommissioned buildings (well matched pair or large sample size recommended).	Simple	Many benefits are first year. Persistence not addressed.

Study/Author	Costs		Energy Benefits		Non-energy Benefits (NEBs)		Persistence Assumptions	
Mills, Meta-analysis (2004)	Moderate	Cx and EBCx: includes Cx provider fee, Cx coordination costs of other parties Cx: Does not include resolution cost for "quality assurance" findings or cost to fix design flaws EBCx: Includes resolution cost	Moderate	Project documentation may have utilized engineering calculations, models, or pre- and post-consumption measurement to quantify savings. 58 % of EBCx and 28 % of Cx projects verified measures to be implemented.	Moderate	First cost non-energy savings ($), and ongoing labor cost savings (type, labor hours saved), and includes a list of other NEBs (Y/N, $)	Simple (number taken from other studies, which are more complex)	Persistence data collected where available from other studies (LBNL, TAMU)). Used their methodology (see Persistence chapter of this report). For the majority of buildings, persistence or measure life was not addressed.[12]
SBW, Northwest Cx & EBCx (2004)	Complex	Includes incremental fee increases, travel expenses, and resolution costs to each party, as reported by respondents.	Moderate	Used project documentation to identified significant and resolved issues, then used standard engineering calculations or parametric modeling to get savings.	Complex	Dollar value estimated three different ways based on telephone survey data with most conservative figure used.	N/A	
California Commissioning Collaborative's Cxdatabase.com (2004)	Moderate	Includes minor capital improvements as a cost of EBCx.[13] Includes incremental costs to other parties	Simple	Respondent provides info for energy-savings calculations for each finding, not required to perform calculation. Persistence and avoided cost info optional. No standardized calculations for energy savings.	Simple	Respondents asked to identify which benefits they received, and given the option of entering a dollar value for the estimated avoided cost. No standardized calculations for avoided costs.	Not assessed	

---

[12] The fast payback times for Cx measures are most likely shorter than the period of erosion of savings.
[13] Allows owner to specify whether O&M staff participation is a cost or a benefit. Does not include resolution costs for "quality assurance" findings as reported by respondents.

Study/Author		Costs		Energy Benefits		Non-energy Benefits (NEBs)		Persistence Assumptions
Moore et al. (2008)	n/a	Only provides payback based on owner's costs.	Complex	Complex spreadsheet calculations, quality-checked by utility program administrators	n/a		n/a	
PECI & Summit Building Engineering (2008)	Complex	EBCx provider costs and implementation costs included.	Complex	Based on utility programs' data, which uses detailed energy calculations and a rigorous review	n/a		n/a	

2.1.4. Recommendations and Decision Points

Table 2.10 displays the 12 diverse studies reviewed in this report. The studies represent a large range of data collection, costs, energy benefit and non-energy benefit methodologies. Although this makes generalizations difficult, their collective efforts point to several recommendations moving forward, and several decision points to which any new study must attend.

Recommendations

> **1. Building commissioning data should be greatly diversified.** In the majority of these studies, building information comes from only a few sources, like a handful of commissioning providers or a large university research department. It is thus unclear how well the findings of these studies will apply to the worldwide commissioning industry. Moving forward, an attempt should be made to gather building data from a much broader base. To date, it has been difficult to collect data from diverse projects because owners do not tend to ask for (or pay for) this kind of data on their own projects, and commissioning providers therefore do not gather it. Collecting commissioning data in a consistent way requires artificial injection of a research project or program to help standardize the way data is gathered and reported by market actors.

> **2. A complex cost-benefit methodology may require continuous data collection throughout the commissioning project, extensive interviews, or both, to acquire a sufficiently detailed reporting of costs.** The data required for complex, and sometimes even moderate, cost methodologies will be difficult to obtain with a retroactive data collection methodology relying solely on documentation. It is nearly impossible to determine from documentation costs that are not explicitly defined during the project. For example, a study may want to include in its cost calculation the cost to the contractor of coordinating with the commissioning provider. If this cost is not defined either during or immediately after the commissioning process it will not be included in documentation (although it may be obtained through a timely interview). As a result, retroactive studies relying mostly on project documentation are often forced to "take what they can get," a methodology which does not lead itself to a consistent definition of commissioning costs. A study employing a complex cost methodology should facilitate accurate data entry by using a collection methodology that allows easy entry of data and thus helps avoid the need for retrospective information-gathering based on project documentation.

> **3. Data validation is an important aspect of any cost-benefit methodology**. In cases where data is not verified, the accuracy of the cost-benefit results may be at risk.

Decision Points in Creating a Cx/EBCx Cost-Benefit Methodology

Creating an appropriate and feasible commissioning cost-benefit methodology that achieves the goals of the project requires careful planning around some key decision points. Ultimately, these decisions lead to a methodology that can have a range of levels of effort to collect and analyze data, as is shown in this chapter. The following key decision points emerged during this analysis of cost-benefit methodologies.

General

- What is the goal of data collection and who is the expected audience?
- What data and formats are appropriate to the study goals and audience?
- What resources do researchers have available to them? This includes both financial resources as well as current and potential data sources.
- How important is verification of data? (Possible levels: reasonableness check, oversight of energy and non-energy benefits calculations, and verification for persistence.)

Data Collection

- Will the study be a one-time event that looks retrospectively at past projects, or will data collection and analysis occur continuously with current and future projects?

Costs

- Should the cost to resolve problems identified by the commissioning provider be counted as a cost of commissioning? If these resolutions are major design changes, should they be counted as a cost of commissioning?
- Should the commissioning-related costs of designers, contractors, and operating staff be counted as costs of commissioning?
- Are tasks performed by a commissioning provider that is out of the scope of commissioning counted as a cost of commissioning? For example, designers are generally tasked with developing the design intent documents. If the designer does not complete these documents, but the commissioning provider must have a complete set to functionally test the systems against, then often the commissioning provider will complete the task.
- Are costs treated differently for new construction commissioning and existing building commissioning?

Energy Benefits

- Will the methodology be whole building or measure-based?
- Will the methodology require monitored data or rely on calculations, and will calculations be validated by monitored data?
- How will measured data be collected (e.g., utility bills, dataloggers, or trends from the building automation system)?
- How will it be tracked that identified measures are implemented?
- Will standardized calculations be used, or guidelines for calculations or modeling?
- What standardized documentation must be collected to support modeling or calculation?
- How will persistence of savings be estimated or verified?

Non-Energy Benefits

- Will an attempt be made to quantify the financial consequences of non-energy benefits?
- If not, how will non-energy benefits be reported and verified?
- If so, will the financial non-energy benefit be self-reported, or will a verification methodology be employed?

2.2 References

Altwies, Joy and Ian B. D. McIntosh. 2001. "Quantifying the Cost Benefits of Commissioning" in Proceedings of the 9th National Conference on Building Commissioning. PECI.

Friedman, Hannah, Tudi Haasl, and Ken Gillespie. 2004. "Creating California's Online Commissioning Case Study Database: Case Studies Go High Tech" " in Proceedings of the ACEEE 2004 Summer Study on Energy Efficiency in Buildings, 3-64 – 3-75, Washington, DC: ACEEE.

Gregerson, Joan. 1997. Commissioning Existing Buildings. E Source TU-97-3. Boulder, CO: E Source, Inc.

Haasl, Tudi, K. Stum, and M. Arney. 1996. "Better Buildings through Improved O&M – A Five Building Case Study." in Proceedings of the 5th National Conference on Building Commissioning.

Heinemeier, Kristin, Michael Martin, Balaji Santhanakrishnan, Anita Ledbetter, Jim Shoop, Wes Harvey, Joseph Martin, and Frank Thomas. 2004. "Commissioning of New Schools: A State Funded Study of Costs and Benefits" in Proceedings of the ACEEE 2004 Summer Study on Energy Efficiency in Buildings, 3-88 -3-99, Washington, DC: ACEEE.

Mills, Evan, Hannah Friedman, Tehesia Powell, Norman Bourassa, David Claridge, Tudi Haasl and Mary Ann Piette. 2004. "The Cost-Effectiveness of Commercial-Building Commissioning." LBNL – 56637, December, 2004.

Moore, Emily, 2008. "California Retrocommissioning Programs: Lessons Learned", Proceedings of the 14th biennial ACEEE Summer Study on Energy Efficiency in Buildings, Pacific Grove, California.

Piette, Mary Ann, B. Nordman, and S. Greenberg. 1995. "Commissioning of Energy-Efficiency Measures: Costs and Benefits for 16 Buildings." LBNL – 36448, April.

Piette, Mary Ann and Bruce Nordman. 1995. "Energy and Cost Savings from Commissioning Efficiency Measures in 16 Buildings" in Proceedings of the 4th National Conference on Building Commissioning.

Portland Energy Conservation, Inc. (PECI). 1997. "What Can Commissioning Do for Your Building?"

PECI. 1996. "Building Systems Deficiency Database." Report to the U.S. Environmental Protection Agency/U.S. Department of Energy under Assistance Agreement CX822837-01-0, December, 1996.

PECI and Summit Building Engineering, "2007 California RCx Market Characterization", Report prepared for the California Commissioning Collaborative.

SBW Consulting. 2004. "Cost-Benefit Analysis for the Commissioning in Public Buildings Project," submitted to the Northwest Energy Efficiency Alliance, posted May, 2004.

Tso, Bing, Marc Schuldt, and Jun Quan. 2004. "The Costs and Benefits of Achieving Silver LEED for Two Seattle Municipal Buildings" in Proceedings of the ACEEE 2004 Summer Study on Energy Efficiency in Buildings, 3-338 -3-350, Washington, DC: ACEEE.

Tso, Bing, Michael Baker, and Andrzej Pekalski. 2002. "The Costs and Benefits of Commissioning Oregon Public Buildings" in: Proceedings of the ACEEE 2002 Summer Study on Energy Efficiency in Buildings, 3.373-3.384, Washington, DC: ACEEE. An earlier version is also included, from the Proceedings of the 8th National Conference on Building Commissioning, 2000. PECI. (Based on SBW Consulting. "Cost-Benefit Analysis for the Commissioning in Public Buildings Project")

Stum, Karl, Tudi Haasl and Dan Krebs. 1994. "Costs and Savings of ECM Commissioning and Inspection – Case Studies from One Utility" in Proceedings of the 2nd National Conference on Building Commissioning, PECI.

Additional Cost-Benefit Studies

Bicknell, Charles and Liza Skumatz. 2004. "Non-Energy Benefits (NEBs) in the Commercial Sector: Results from Hundreds of Buildings" in Proceedings of the ACEEE 2004 Summer Study on Energy Efficiency in Buildings, 4-10 – 4-22, Washington, DC: ACEEE.

Bushell, Christopher, Jack Wolpert, Mandeep Singh, and William Burns. 2002. "Retro Commissioning at Chicago Public Schools" in Proceedings of the 10th National Conference on Building Commissioning, PECI.

Chao, Mark and David B. Goldstein. 1998. "Commissioning, Energy Efficiency, and the Asset Value of Commercial Properties: Perspectives from Appraisers and Financial Stakes for Owners" in Proceedings of the 6th National Conference on Building Commissioning, PECI.

Claridge, David E., Charles H. Culp, Song Deng, W.D. Turner, J.S. Haberl, and Mingsheng Liu. 2000. "Continuous Commissioning of a University Campus" in Proceedings of the 8th National Conference on Building Commissioning, PECI.

Cohan, David and Phil Willems. 2001. "Construction Costs and Commissioning" in Proceedings of the 9th National Conference on Building Commissioning, PECI.

Coleman, James D. 1998. "Three Building Tune-Up Case Studies" in Proceedings of the 6th National Conference on Building Commissioning, PECI.

Cox, Robert and April Williams. 2000. "Quantifying Costs & Benefits of Commissioning" in Proceedings of the 8th National Conference on Building Commissioning, PECI.

Deall, Jerry H., Jack S. Wolpert, Mandeep Singh, and James Kelley. 2002. "Savings Due to Commissioning at the King Center, (Performing Arts Building)" in Proceedings of the 10th National Conference on Building Commissioning, PECI.

Dorgan, Chad, Robert Cox, and Charles Dorgan. 2002. "The Value of the Commissioning Process: Costs and Benefits." The Austin Papers: Best of the 2002 International Green Building Conference, Compiled by Environmental Building News, pp. 25-30.

Khan, Aleisha. 2002. "Retrocommissioning of Two Long-Term Care Facilities in California. Institute for Market Transformation 2002, available online at www.imt.org/papers.

Kumar, Satish and William J. Fisk. 2002. "IEQ and the Impact on Building Occupants" in the ASHRAE Journal.

Kumar, Satish and William J. Fisk. 2002. "The Role of Emerging Energy-Efficient Technology in Promoting Workplace Productivity and Health: Final Report" Publication source unknown. February 13, 2002.

Martinez, Mark S. 1999. "Energy Efficient Retrofit Commissioning – Tricks of the Trade to Avoid Claims and Unhappy Customers" in Proceedings of the 7th National Conference on Building Commissioning, PECI.

McHugh, Jonathon, Lisa Heschong, Nehemiah Stone, Abby Vogen, Daryl Mills, and Cosimina Panetti. 2002. "Non-Energy Benefits as a Market Transformation Driver" in Proceedings of the ACEEE 2002 Summer Study on Energy Efficiency in Buildings, 6.209-6.220, Washington, DC: ACEEE.

Milton, Donald K. 2002. "IEQ and the Impact on Employee Sick Leave" in the ASHRAE Journal, 2002.

Nelson, Norman L. 1999. "Avoiding Litigation through Proper Commissioning: My Experience as an Expert Witness" in Proceedings of the 7th National Conference on Building Commissioning, PECI.

Nicholls, Andrew, Sean McDonald, and John D. Ryan. 2002. "Estimating the Benefits from Building Technologies: Issues, Challenges, and Lessons Learned from Blue Ribbon Panel Reviews" in Proceedings of the ACEEE 2002 Summer Study on Energy Efficiency in Buildings, 10.221-10.236, Washington, DC: ACEEE.

Parker, Gretchen and Mark Chao. 1999. "Management and Documentation of Building Energy Performance by Real Estate Investment Trusts" in Proceedings of the 7th National Conference on Building Commissioning, PECI.

Parker, Gretchen, Mark Chao, and Ken Gillespie. 2000. "Energy-related Practices and Investment Criteria of Corporate Decision-Makers" in Proceedings of the ACEEE 2000 Summer Study on Energy Efficiency in Buildings. Washington, DC: ACEEE.

Parks, Jim, Mazin Kellow, Debby Dodds, and Greg Cunningham. 2001. "Getting the Most Out of Your Building: SMUD's Commissioning Program" in Proceedings of the ACEEE 2000 Summer Study on Energy Efficiency in Buildings, 4.247 - 4.256, Washington, DC: ACEEE; also published as "SMUD's Retrocommissioning Program: Improving With Time" in Proceedings of the 9th National Conference on Building Commissioning, PECI.

Tseng, Paul. 1998. "Building Commissioning: Benefits and Costs." Heating/Piping/Air Conditioning, April, 1998, pp. 51-59.

Veltri, Anthony. 2002. "Development of an Integrated Commissioning Strategy and Cost Model" in Proceedings of the 10th National Conference on Building Commissioning, PECI.

Wei, Guanghua, Mingsheng Liu, Martha J. Hewett, and Mark W. Hancock. 2001. "Continuous Commissioning of a Hospital Complex" in Proceedings of the 9th National Conference on Building Commissioning, PECI.

Wilkinson, Ronald J. 2000. "Establishing Commissioning Fees" ASHRAE Journal. February 2000, pp. 41-51.

3. COMMISSIONING PERSISTENCE

3.1 Methodologies for Determining Persistence of Commissioning Benefits

3.1.1 Commissioning Benefits

The benefits of Cx and EBCx are normally verified as part of the project, whether they are energy-related benefits or non-energy benefits (NEBs). How those benefits last over time is a subject of much study and discussion, in two principle areas:

- How will certain Cx/EBCx measure types persist over time (typically measured as percentage degradation over certain time periods in years),
- The robustness of methods used to ascertain measure persistence.

The one-time, or inherently persistent benefits normally reduce construction costs directly or indirectly. Table 3.1 lists a number of reported benefits of commissioning (Mills et al. 2005, Friedman et al. 2002, Liu et al., 2002) that appear to generally fall in the category of inherently persistent benefits. They have been grouped as design benefits, construction benefits, early occupancy benefits, and "other", primarily based on when they occur in the design/construction process. The benefit from design improvements inherently occurs once, but this benefit persists until the building is renovated or equipment fails and is replaced. Many more design benefits than those listed may result from commissioning. The benefits that speed up or make the construction process flow more smoothly will clearly provide a one-time benefit. The benefits that make early occupancy a more seamless process will generally be one-time benefits, though the items related to safety and liability may be viewed as on-going benefits.

The role of commissioning in qualifying a building for a Leadership in Energy and Environmental Design (LEED) rating or participation in a utility program may provide long-term benefits, but are treated as inherently persisting. A thorough EBCx process can be a significant enabling factor for a thorough building retrofit.

Whether viewed as one-time benefits, or as longer term, the commissioning benefits shown in Table 3.1 will not be considered among those that may degrade over time. Hence they will be assumed to be inherently persistent in the context of the persistence methodology presented here.

The benefits listed in Table 3.2 have also been reported as commissioning benefits (Mills et al. 2005, Friedman et al. 2002, Liu et al., 2002), but these are items related to the operation of the building that are thought to be more likely to change over time, particularly if they are the result of the implementation of practices that are not widely understood by the community of building operators. Hence these benefits are treated as commissioning benefits that may not persist.

Table 3.1 Inherently persistent benefits of commissioning

Design Benefits
<ul><li>Equipment right-sizing</li><li>Improved equipment layout</li></ul>
Construction Benefits
<ul><li>Improved project schedule</li><li>Clarified delineation of responsibilities among team members</li><li>Fewer change orders</li><li>Less disagreement among contractors</li><li>Reduced contractor call-backs</li><li>More vigilant contractor behavior (knowing that Cx will follow their work)</li><li>Reduced testing and balancing (TAB) costs</li></ul>
Early Occupancy Benefits
<ul><li>Smoother process and turnover</li><li>Less disruption to occupancy and operations during turnover</li><li>Fewer warranty claims</li><li>Improved safety</li><li>Reduced liability</li></ul>
Other
<ul><li>Compliance with LEED or other sustainability rating system</li><li>Qualification for rebate, financing or other services</li><li>Qualification for participation in utility program</li><li>An enabling factor for comprehensive system overhaul</li></ul>

Table 3.2 Commissioning benefits that may not persist

Easier to Quantify
Reduced energy consumption
Ensured or improve indoor environment /occupant comfort
Ensured adequate indoor air quality
Improved water utilization
More Difficult to Quantify
Repaired or accelerate repair of a problem
Avoided premature equipment failure
Reduced operations and maintenance costs
Increased occupant productivity
Improved documentation
Increased in-house staff skills, knowledge, awareness
Improved operational efficacy
Provided sustainable engineering solutions to operational problems
Ensured proper system performance (energy and non-energy systems)

3.1.2 Measures of Benefit Persistence

The first four items shown in Table 3.2 are commissioning benefits that may be quantified if suitable baselines for comparison are available. The *International Performance Measurement and Verification Protocol* (IPMVP 2001) is widely used to determine savings in energy and water resulting from either retrofits or operational changes. It provides procedures that may also be applied to new buildings if the impact of commissioning measures implemented can be accurately assessed in a simulation. Comfort has been widely studied, and measures of comfort such as dry bulb temperature and relative humidity can be measured and logged. Likewise, CO_2 and other measures of indoor air quality may be measured. It is assumed that new buildings will provide comfort and quality indoor air, so it will be difficult to document commissioning benefits to comfort or indoor air quality in new buildings. However, when commissioning is carried out in an existing building, these changes can be documented with appropriate measurements before and after commissioning. These measurements are most likely to be made if a serious comfort and/or air quality problem provides a significant part of the motivation for commissioning the building.

The remaining items listed in Table 3.2 are much more difficult to document, beyond the documentation of specific commissioning measures that have been implemented and verification that these measures are still in place months or years later. Hence the only further treatment of these benefits within the proposed methodology will be through documentation of specific measures related to these benefits.

Given this context, and based on a review of the existing literature on persistence of commissioning benefits (Frank et al. 2005), the proposed methodology for determining persistence of commissioning benefits will specifically treat the persistence of the energy, water, comfort, and indoor air quality benefits of commissioning in a quantitative manner. It will treat all other benefits through examination of the persistence of specific commissioning measures that have been implemented.

3.1.3 Persistence of Energy Benefits from Commissioning

The energy benefits of commissioning will initially be determined using an appropriate methodology from the International Performance Measurement and Verification Protocol (IPMVP 2001). This protocol provides a general approach that compares measured energy use or demand before and after implementation of an energy savings program using the equation:

Energy Savings = Base year Energy Use – Post-Retrofit Energy Use ± Adjustments

The "Adjustments" term brings energy use in the two time periods to the same set of conditions by adjusting for differences in weather, occupancy, plant throughput, and equipment operations. These adjustments are made routinely for weather changes, or as needed for occupancy changes, scheduling changes, etc.

Four basic options are presented for determining energy savings within the IPMVP. These options are briefly described in Table 3.3. Within the context of this methodology, the only option that is considered appropriate for determining the energy savings from commissioning of a new building is Option D, Calibrated Simulation. This permits the calibration of a simulation to the measured consumption of the building following commissioning, followed by simulation of the changes made during commissioning. For existing buildings that are commissioned, energy savings from comprehensive commissioning projects may be evaluated using either Options C or D. If the savings from the commissioning process are too small to evaluate in one of these ways or only one or two measures are expected to result in energy savings, then Option B may be appropriate. Option A will rarely be appropriate. The detailed procedures in the protocol are to be used.

Following determination of energy savings in multiple years using the selected procedure, savings from each year in which savings are determined will be further normalized to a common weather year to eliminate bias in the persistence determination from weather differences in the different years. Other adjustments may also be made when warranted by known conditions.

Table 3.3 The four IPMVP energy savings options
Source: IPMVP 2001

M&V Option	How Savings Are Calculated	Typical Applications

3.1.4 Persistence of Water Savings Benefits from Commissioning

The IPMVP methodologies for determining water savings are the same as those used to determine energy savings. In these cases, it becomes important to consider precipitation if the building water consumption includes water uses for exterior landscaping.

3.1.5 Persistence of Commissioning Measures

In some cases, appropriate metering is not installed or baseline information needed to determine energy savings is not available. In other cases, the measures of interest may not impact energy consumption, but may impact other benefits of commissioning as discussed in the section on "Measures of Benefit Persistence." In these cases, persistence shall be determined by comparing a list of documented commissioning measures that were implemented during the commissioning process with the measures that are subsequently documented as being in place or operational during the time when persistence is being checked.

When used to evaluate measures that impact energy consumption, the most comprehensive systematic listing of measures that may be considered is probably that of Mills et al. (2005). They used a matrix that included the specific commissioning measures in the four categories listed in Table 3.4. These measures were then considered as being applied to deficiencies in the areas or systems shown in Table 3.5.

Table 3.4 Specific commissioning measures

Design, Installation, Retrofit, Replacement
• Design change
• Installation modifications
• Retrofit/equipment replacement
• Other
Operations and Control
• Implement advanced reset
• Start/stop (environmentally determined)
• Scheduling (occupancy determined)
• Modify setpoint(s)
• Equipment staging
• Modify sequence of operations
• Loop tuning
• Behavior modification/manual changes to operations
• Other
Maintenance
• Calibration
• Mechanical fix
• Heat transfer maintenance
• Filtration maintenance
• Other
Miscellaneous
• Deficiency unmatched to specific measure

Table 3.5 Areas or systems in which measures correct deficiencies

HVAC (combined heating and cooling)
Cooling plant
Heating plant
Air handling and distribution
Terminal units
Lighting
Envelope
Plug loads
Facility-wide (e.g. EMCS or utility related)
Other
Deficiency unmatched to specific measure

3.2 Review of Literature on the Persistence of Commissioning Benefits in New and Existing Buildings

3.2.1 Introduction

In recent years the topic of persistence of benefits has gained more interest both for existing building and new building commissioning. Several studies have been performed and published examining both aspects of this topic. This review will summarize the key results of these studies. The categories presented are persistence of commissioning measures in existing buildings, persistence of commissioning measures in new buildings, strategies for improving persistence in new and existing buildings, and related reports. This topic is relatively new, and the only relevant projects identified in the literature to date involve a total of 37 buildings:

- 10 Existing Buildings Commissioned at Texas A&M University – Turner et al. (2001) and Claridge et al. (2002, 2004)
- 8 Existing Buildings Commissioned in Sacramento, California – Bourassa et al. (2004)
- 8 Existing Buildings Commissioned in Oregon – Peterson (2005)
- 1 Existing Building Commissioned in Colorado – Selch and Bradford (2005)
- 10 New Buildings Commissioned – Friedman et al. (2002, 2003a, 2003b)

Since the total literature identified consists of published papers and reports from only five projects directly related to persistence, the summaries presented for each project are considerably more detailed than is customary in a literature review.

3.2.2 *Persistence of Commissioning Measures in Existing Buildings*

Texas A&M: Ten Buildings at Texas A&M University

A study was performed in 2000 to evaluate the persistence of savings in ten buildings on a university campus three years after the buildings participated in existing building commissioning (Turner et al. 2001). The objectives of the study were to determine quantitatively how much savings degradation occurred and the major causes of any observed degradation. The investigation did not focus on the detailed measures implemented in each building but rather on the degree to which the measures implemented in the EBCx process had been maintained, as indicated by examination of energy use data, the EBCx reports, and the control settings in place on the main energy management control system.

The study was conducted in five major parts. First, buildings were selected to be studied. Second, savings calculations were performed based on energy usage data from the different periods needed. Third, field examination and commissioning follow-up was conducted on two buildings in which major savings degradation occurred. Fourth, operational and controls changes that could have contributed to changes in building performance after commissioning were identified. And fifth, calibrated simulations of some of the buildings were performed to verify the effects of the identified changes on energy consumption.

A preliminary group of 20 buildings which had been commissioned in 1996 or 1997 was initially selected. An office review of information on the EBCx measures implemented and available information on operating parameters before and after existing building commissioning was then conducted. Based on this review, the ten buildings with the most complete information concerning the EBCx process and energy consumption data were selected. None of the buildings in this group received capital retrofits during the period of 1996 to 2000. Five buildings were commissioned in 1996 and the other five were finished in 1997. In each of these buildings, commissioning measures were identified by the EBCx provider and then implemented by the provider, after receiving the concurrence of the building owner's representative. Since all ten buildings were located on a university campus, they primarily consisted of classrooms, laboratories, and offices, with one volleyball arena.

The energy usage data for these buildings had been monitored and was obtained beginning with the period shortly before EBCx and ending in 2000 when the study was performed. For comparison purposes, all of the energy data was normalized to a single year of weather data. Because the weather data for the year 1995 most closely approximated average weather conditions for the years studied, it was chosen as the baseline year. Energy use before and after the EBCx process were compared. In this study savings from the EBCx process were determined by using Option C of the International Performance Measurement and Verification Protocol, which determines savings using measured energy use at the whole facility level. This required that baseline models of the consumption be formulated for each major source of energy use in each building. Chilled water and hot water energy consumption were measured for each year, and three-parameter or four-parameter change-point models of cooling and heating consumption were determined as functions of ambient temperature using a modeling program.

The process of calculating the yearly savings required the development of five separate chilled water models and five hot water models for each building, one for each year, including the baseline model. The consumption and savings for each year were then normalized to 1995 weather by using the models for each year's data with the 1995 temperature data to determine the savings for each year. Electricity savings were determined without normalization since the buildings did not have chillers, and electricity consumption is not appreciably affected by ambient temperature.

Follow-up was performed on two buildings with significant savings degradation. This was done primarily through a field investigation of the buildings to determine what changes had occurred that would produce the changes. Equipment performance and Energy Management Control System (EMCS) control settings were examined to evaluate possible causes for degradation.

Information was then gathered on controls and operational changes that had occurred in the buildings during the period studied. This was done by examining the EBCx reports and interviewing the engineers and maintenance personnel who had responsibility for each building. These interviews provided identifiable reasons for many of the changes in savings seen in the buildings.

In order to quantify the effect of each operational or control change identified, it was decided that the energy usage of the buildings would be modeled using a computer simulation program. The rough simulations would then be calibrated until they provided accurate representations of the actual energy use. These simulations would then demonstrate how much of an effect each control or operational change had on the building energy use.

Texas A&M: Results

All ten buildings showed significantly reduced chilled water and hot water energy consumption since EBCx, although the savings generally decreased somewhat with time. Eight buildings had larger HW savings in 1998 than in 1997 as a consequence of hot water loop optimization conducted in 1997 and final EBCx actions. Overall the electricity consumption remained fairly constant, with three buildings showing small increases in consumption (negative savings). The average electricity savings for the ten buildings from 1997 to 2000 were 10.8 %. Figure 3.1 and Figure 3.2 show the chilled water and hot water savings trends for the years following the building EBCx.

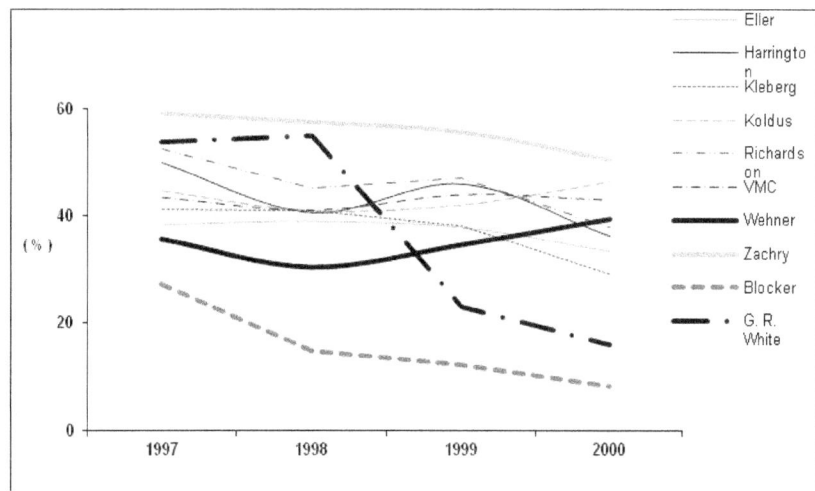

Figure 3.1 Chilled water savings persistence after EBCx.

Overall, chilled water savings for the three years following EBCx averaged 39.3 % of the pre-commissioning baseline. Eight of the buildings showed good persistence of savings for chilled water (less than 15 % change during the three to four years after EBCx), while the other two displayed significant degradation. The Blocker building had 19 % degradation, and the G. R. White Coliseum had a dramatic savings degradation of 38 %.

Hot water consumption was reduced significantly in the years following EBCx, but the savings fluctuated widely from year to year. Savings increased from 1997 to 1998 in most buildings due to optimization in the hot water loop in 1997 and some ongoing EBCx work. The ten buildings averaged hot water savings of 65.0 % after EBCx.

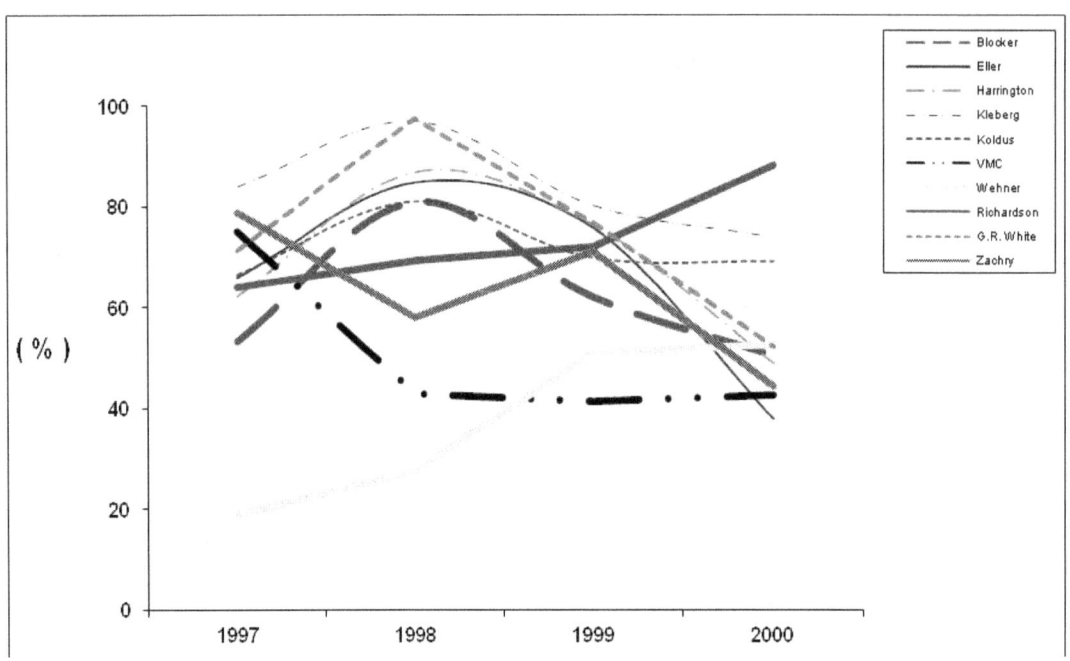

Figure 3.2 Hot water savings persistence after EBCx.

Based on the historic campus energy costs of $4.42/GJ ($4.67/MMBtu) for chilled water, $4.50/GJ ($4.75/MMBtu) for hot water, and $0.02788/kWh for electricity, the cumulative savings from EBCx in these ten buildings were $4 439 000 for the period of 1997 to 2000. Only three buildings had year 2000 savings greater than 1998 savings, and the increase in two of these was approximately 2 % of baseline consumption, which is well within the range of normal year-to-year variation. The savings of the other buildings decreased.

Follow-up investigations of the two buildings with significant savings degradations revealed serious equipment malfunction and controls failure. In the Kleberg building, two chilled water control valves were found to be leaking badly, and combined with a failed electronic to pneumatic switch and high water pressure, caused low discharge temperatures and continuous reheat operation. In addition, failed sensors caused the outside air dampers to remain fully open, and leaking damper actuators in a number of VAV boxes resulted in simultaneous heating and cooling. The G. R. White Coliseum was found to have a controls malfunction resulting in simultaneous heating and cooling, with two of the thirteen air handling units operating in heating mode, while the rest operated in cooling mode. These equipment and controls problems in these two buildings were the primary causes of the savings degradation observed. Because these problems did not result in comfort problems in the buildings, they may have gone undetected had the energy consumption not been monitored and compared with previous data.

The energy management control system settings were evaluated for the buildings to determine why the changes in savings occurred. Three major control settings were examined: cold deck or cooling discharge temperatures, hot deck temperatures, and static pressure settings. The cold deck or cooling coil discharge temperatures were reset during EBCx to save chilled water consumption. It was found that for eight of the ten buildings in 2000, the temperatures had been lowered and were requiring more cooling. This led to chilled water savings degradation, particularly in the Blocker building. Five of the ten buildings had dual duct systems, and of these five, three of the hot deck temperature set points were at different values in 2000 than they had been upon completion of EBCx. This resulted in more hot water consumption. Static pressure set points affected chilled water, hot water, and electricity consumption. Of the nine buildings with variable air volume systems, only one (Koldus) still had the same static pressure set point in 2000 that it had been set to during EBCx. The other buildings were requiring more static pressure, and therefore using more energy. It is worth noting that the Koldus building showed no serious savings degradation of any kind in this study.

Data were gathered from engineers and maintenance personnel to attempt to verify the controls changes and explain them. It was found that the G. R. White Coliseum, which saw significant savings degradation in chilled water and hot water savings, had experienced malfunctions in air handling unit controls that caused simultaneous heating and cooling to occur throughout the year. Almost all of the savings degradation for this building could safely be attributed to these problems. It was also found that the Kleberg building had experienced some significant equipment problems that could explain some of the degradation in savings that occurred. No other building was reported to have experienced equipment problems of the same caliber as these two cases, but controls changes in the other buildings were verified through investigation. With the assembly of this type of information, simulated calibrations could be made for the buildings. Lack of data and other problems such as the one mentioned for the G. R. White Coliseum. G.R. White allowed only five of the ten buildings to be simulated. Three simulations were performed for each of these buildings, one for the pre-commissioning period, one for the year after EBCx, and one for the year 2000. Factors considered in the simulations included control settings changes, operator overrides on the controls, and physical changes in the system such as broken or repaired valves, sensors, etc. Detailed simulations of the control changes in Eller O&M, Harrington Tower, VMC Addition and Wehner showed that the RMS difference between the changes observed between the post-commissioning periods and year 2000 was only 1.1 %, suggesting that the changes in savings for these buildings were almost entirely due to the control changes identified.

Overall, equipment malfunction and changes made in cold deck and hot deck temperature settings following EBCx were the major reasons for changes in chilled water and hot water energy consumption and savings after EBCx.

Table 3.6 is a summary of the money saved in the year 1998 as compared with the money saved in the year 2000 for each of the ten buildings examined.

Table 3.6 Cost savings calculations for the year 1998 and the Year 2000.

No.	Buildings	Type	Baseline Energy Use (GJ/y) (MWh/y)	Year 1998 Savings				Year 2000 Savings
				Energy Use (GJ/y) (MWh/y)	Savings GJy MWh/y	Cost Savings Each US$/y	Total US$/y	Total US$/y
1	Blocker	CHW	24 218	20 605	3 613	$ 15 993		
		HW	9 216	1 768	7 448	$ 33 533		
		Elec.	4 832	3 883	950	$ 26 477	$ 76 003	$ 56 738
2	Eller O&M	CHW	32 311	19 687	12 623	$ 55 875		
		HW	8 001	1 218	6 783	$ 30 539		
		Elec.	4 891	3 675	1 217	$ 33 925	$ 120 339	$ 89 934
3	G.R. White Coliseum	CHW	19 911	8 979	10 932	$ 48 386		
		HW	22 319	580	21 740	$ 97 875		
		Elec.	1 480	1 168	312	$ 8 712	$ 154 973	$ 71 809
4	Harrington Tower	CHW	14 959	8 883	6 076	$ 26 895		
		HW	7 276	964	6 311	$ 28 413		
		Elec.	1 666	1 336	330	$ 9 189	$ 64 498	$ 48 816
5	Kleberg Building	CHW	62 534	36 894	25 640	$ 113 491		
		HW	43 059	1 281	41 777	$ 188 086		
		Elec.	5 511	5 067	444	$ 12 380	$ 313 958	$ 247 415
6	Koldus Building	CHW	23 173	13 703	9 470	$ 41 916		
		HW	2 218	421	1 798	$ 8 093		
		Elec.	2 850	2 597	253	$ 7 067	$ 57 076	$ 61 540
7	Richardson Petroleum	CHW	30 096	16 497	13 599	$ 60 191		
		HW	19 230	5 895	13 335	$ 60 035		
		Elec.	1 933	1 914	19	$ 519	$ 120 745	$ 120 666

No.	Buildings	Type	Baseline Energy Use (GJ/y) (MWh/y)	Year 1998 Savings				Year 2000 Savings
				Energy Use (GJ/y) (MWh/y)	Savings GJy MWh/y	Cost Savings Each US$/y	Savings GJy MWh/y	Total US$/y
8	VMC Addition	CHW	43 143	25 406	17 738	$ 78 513		
		HW	3 766	2 153	1 613	$ 7 260		
		Elec.	4 186	4 140	46	$ 1 286	$ 87 059	$ 92 942
9	Wehner CBA	CHW	20 249	14 073	6 177	$ 27 339		
		HW	14 130	10 250	3 880	$ 17 469		
		Elec.	2 555	2 446	109	$ 3 026	$ 47 834	$ 68 145
10	Zachry Engr. Center	CHW	43 071	18 334	24 738	$ 109 496		
		HW	8 098	3 408	4 690	$ 21 114		
		Elec.	7 502	6 793	710	$ 19 789	$ 150 400	$ 127 620

Type					Totals	Year 1998	Year 2000
Chilled Water		313 666	183 062	130 605	$ 578 096		
Hot Water		137 314	27 940	109 374	$ 492 417		
Electricity		37 407	33 018	4 389	$ 122 371	$1 192 884	$ 985 626

* The baseline energy use data for two buildings were created based on the average savings of the other buildings because they did not have enough data.

**To obtain MMBtu/yr, multiply the number of GJ/yr by 0.9478.

Texas A&M: Conclusions

Table 3.7 summarizes the savings history of this group of ten buildings. The savings in 1998 following initial retro commissioning corresponded to average energy cost savings of 39 % for the ten buildings. Savings decreased to 32.3 % over the next two years – still a highly significant level of savings.

Table 3.7 Summary of savings history in ten retro-commissioned buildings at Texas A&M

	Baseline use (US$/y)	1998 Cx savings (US$/y)	Persistence of savings in 2000 (US$/y)
10 Buildings	$3 049 487	$1 192 000 (39.1 %)	$984 516 (32.3 %)
8 Buildings	$2 195 307	$723 376 (32.9 %)	$666 108 (30.3 %)
2 Buildings	$854 180	$468 624 (55 %)	$314 408 (37 %)

Investigation showed that two of the buildings, G. R. White Coliseum and Kleberg, accounted for 3/4 of the total savings degradation, and both had experienced major equipment and controls malfunctions which were the primary causes of their degradation. Following correction of these problems, savings were restored to earlier levels. In the remaining eight buildings, savings changes were rather small, declining from 32.9 % to 30.3 % in aggregate.

All but one of eight buildings had experienced at least some changes in EMCS control settings. To verify the impact of the EMCS changes on energy consumption, the calibrated simulation process was performed on the four buildings with the most complete data sets. Simulation was conducted for a pre-commissioning period, a post-commissioning period soon after EBCx and for the year 2000 for each building. It was found that the changes in consumption observed following EBCx in these buildings were consistent with those due to the identified controls changes, with an RMS difference of only 1.1 %. Control changes accounted for the savings increase observed in the Wehner Building as well as the decreases observed in the other three buildings. This suggests that the changes in savings these four were almost entirely due to the control changes.

Based on the results of this study of ten buildings, it was concluded that:
- Basic existing building commissioning measures are quite stable,
- Savings should be monitored to determine the need for follow-up, and
- Steps should be taken to inform operators of the impact of planned/implemented control changes.

SMUD: Eight Buildings in SMUD Program in Sacramento

In 2003, a study was performed by the Lawrence Berkley National Laboratory (LBNL) on eight buildings that had undergone EBCx through the Sacramento Municipal Utility District (SMUD) EBCx program (Bourassa et al. 2004). The objective of the study was to determine the extent to which EBCx measures were implemented, and the magnitude and persistence of energy savings achieved. Another objective was to see if the two primary goals of the SMUD EBCx program had been met: reduced overall annual building energy consumption, and improved energy efficiency awareness and focus in the customer. The eight buildings selected for the study consisted of six office buildings, one laboratory, and one hospital. Four of the buildings participated in EBCx in 1999, and the other four in 2000. In this program, the EBCx provider worked with the building operators to develop the recommended measures. The measures selected for adoption were subsequently implemented by the building staff and/or contractors over a period of up to two years.

SMUD: Energy Analysis

The energy savings obtained in the years following EBCx were determined and compared. In order to be able to compare energy savings in the different buildings over the years examined, baseline energy consumption was established for each building based on pre-EBCx energy use. Electricity use data were collected from monthly utility bills for each building. Four buildings also had metered data recorded at 15 minute intervals. Gaps in utility bills were filled from site records or regression analysis.

The energy consumption data were normalized to a common weather year and to a common billing cycle of 30.5 d. The savings were calculated using spreadsheets, based on the normalized data, which allowed for a simpler and more robust statistical comparison. Another set of savings was also calculated, based on the EBCx report predictions. Adjustments were made for a capital retrofit in one of the buildings. The cost of EBCx was also estimated for each of the buildings, based on three categories: SMUD's EBCx costs, the site's EBCx costs, and the EBCx measure implementation costs. Based on the estimated costs and savings, simple payback periods for EBCx at each of the sites were calculated and compared.

The electrical savings observed for each building over the years following EBCx are shown in Fig. 3.3

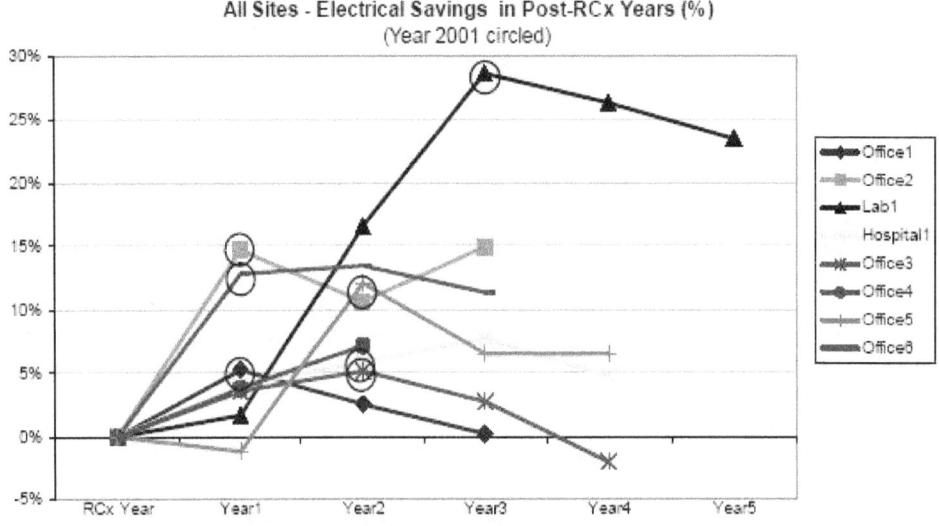

Figure 3.3 Electrical savings following EBCx for each of the buildings.

The aggregate savings for the sites are shown in Fig. 3.4. The buildings are grouped together according to the number of years of data available after EBCx. Note that the "two-year" line includes data for all buildings for the first two years following EBCx, that the "three-year" line includes data for all buildings having three years of data following EBCx, and the "four-year" line includes data for all buildings having four or more years of data following EBCx. Comparison with the data in Fig. 3.3 suggests that the peak in year 3 may be largely due to the one building whose savings peaked in year 3.

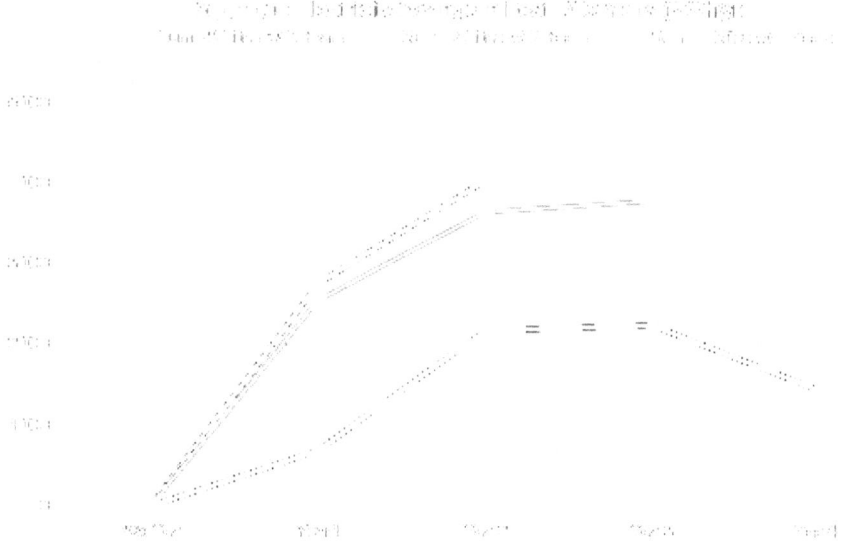

Figure 3.4 Plot of aggregate post-EBCx electricity savings.

These plots demonstrate the observed trend in energy savings for the commissioned buildings. During the first two years the savings generally increased. This was expected because of the length of time needed for the EBCx measures to be implemented. In the third year the savings began to level off, and the fourth year generally showed a declination in the electricity savings. A comparison with the predicted savings estimated in the EBCx reports revealed that on average these reports underestimated the savings by 27.5 %.

The average electricity savings for all the sites over all the years was 7.3 % per year. Natural gas usage could only be obtained for four of the buildings. The savings for natural gas were considerably lower, but since Sacramento is dominated by cooling needs, the lower natural gas savings only reduced the average total energy savings in these four buildings to 6.1 % per year.

The payback periods for the EBCx projects all proved to be attractive, with the longest period being 2.3 years. Table 3.8 lists the estimated costs, annual savings, and payback period for each site, as well as a price per square foot of the building.

Table 3.8 Costs, energy savings, and payback periods for the eight sites studied.

Building	EBCx study costs (agent cost 25k, balance incurred by site)	Estimated measure implmentation costs	Predicted average annual savings ($)	Post EBCx average annual savings ($)	Predicted simple pay-back	Post EBCx simple pay-back	EBCx study costs ($/m²)	EBCx study & implement. costs ($/m²)
Office 1	$28 000	$1 710	$24 500	$13 000	1,2	2,3	2,05	2,15
Office 2	$26 500	$20 500	$21 900	$27 900	2,1	1,7	0,75	1,29
Lab 1	$26 000	$12 370	$64 800	$40 100	0,6	1,0	3,01	4,41
Hospital	$28 300	$11 180	$35 200	$30 900	1,1	1,3	1,18	1,61
Office 3	$25 400	$150	$6 400	$22 400	4,0	1,1	0,65	0,65
Office 4	$26 817	$8 380	$8 400	$22 600	4,3	1,6	0,86	1,18
Office 5	$26 817	$4 350	$9 100	$15 800	3,4	2,0	0,86	0,97
Office 6	$26 700	$3 000	$11 200	$48 600	2,7	0,6	0,97	1,08
All sites total	$214 533	$61 650	$181 600	$221 200	1,5	1,2	0,97	1,29

*To obtain costs in $/ft², multiply the number of $/m² by 0.0929.

SMUD: Measure Persistence Analysis

A series of interviews and site visits were used to determine the persistence in the EBCx measures recommended. The eight EBCx reports recommended a total of 81 corrective measures, of which 48 were implemented. Of these 48, it was found that 81 % had persisted, in that they were still in effect at the time of the study. It was discovered that four of the measures had been abandoned completely, all of which were air distribution component recommendations. Five of the measures had undergone evolution by the building engineers because the original measures had not resolved the problems.

Surveys were given at the sites to determine attitudes regarding the EBCx process, as well as its benefits. All of the sites reported that EBCx was a worthy process. Four of the sites listed training as the primary non-energy benefit from EBCx. The most cited downside to EBCx was the time intensive nature of the process. All of the sites came out of the EBCx process with ideas on how to retain the commissioning benefits over time, the most common solutions being preventative maintenance plans. All of the sites would undertake EBCx again, but only two had potential internal funding.

SMUD: Conclusions

Some important EBCx process factors that this study identified were:

- The commissioning authority is most effective when he is both an expert and a teacher.
- Building engineers prefer to evolve the settings on a recommendation that doesn't work, rather than revert to the previous condition.
- EBCx appears to raise energy efficiency awareness.
- EBCx funds are constrained within building management budgets.

The energy analysis results showed:

- Analyses should not emphasize first-year savings because savings typically take two to three years to fully manifest.
- Energy savings persist to four years or more, although some degradation begins in the third year.
- The EBCx energy savings predictions were reasonably accurate.
- Building managers lack tools for tracking energy performance.
- EBCx cost pay back was shorter than the apparent savings persistence.
- EBCx focused mostly on electricity savings and some natural gas tradeoffs in the savings occurred.

The study suggested several recommendations for the SMUD EBCx program:

- Develop measure implementation tracking agreements, possibly with inspections.
- Explore methods to conduct a three year post-EBCx energy consumption analysis using the billing history.
- Develop simple Performance Tracking Tools for the building operators.
- Develop an extension to the program whereby participants are eligible for new incentives in year 4 to evaluate and update the EBCx as necessary.

On the whole, the SMUD EBCx program's two broad goals were met at the eight sites. Aggregate post-EBCx savings were strong, peaking at approximately 4 420 MWh and the program helped educate site staff about energy efficiency and the role operations and maintenance plays.

Oregon: Oregon Case Study

A study performed in Oregon in 2004 examined eight Intel buildings that had been retro-commissioned in 1999 and 2000 (Peterson 2005). The buildings were located on the Intel Jones Farm and Hawthorn Farms campuses. Kaplan Engineering and PECI performed the EBCx for these buildings through funding from Portland General Electric (PGE). At the time EBCx occurred, it was estimated that electricity savings of nearly 3.5 million kWh annually would result from the low cost energy efficiency measures (EEMs) proposed. The purpose of this study was to examine the energy usage of the buildings to determine what percentage of the original savings was still being achieved four years later. At the same time, it was desired to determine how many of the EEMs proposed were still being utilized.

Three of the buildings studied were located on the Hawthorn Farms Campus (HF), and were designated HF1, 2, and 3. The buildings combined for a total of 59 457 m^2 (640 000 ft^2), and were served by a central chiller and boiler plant. HF1 had Direct Digital Controls (DDC) interfaced with pneumatic actuators, and the other two buildings were upgraded to DDC control in 2000. The remaining five buildings studied were located on the Jones Farm Campus (JF). They combined for a total of 130 063 m^2 (1 400 000 ft^2), with over 40 major air handling systems served by two central chiller plants and two hot water boiler plants. Most of the spaces on both campuses were served by variable air volume (VAV) systems.

Three reports generated at the time of EBCx were examined to determine what measures had been implemented. The current status of these measures was determined through random sampling, with functional testing or trending being used as appropriate. For HF1, the terminal reheat units were serviced at the time of EBCx to ensure proper damper motion. At the time of this study, random sampling discovered no noticeable damper movement from full cooling to full heating in 60 % of the units. The savings for this measure did not persist, probably due to the aging pneumatic system. For HF 1, 2, and 3, EBCx had modified outside air intake controls to allow for the economizing cycle to function. At the time of the study, random sampling revealed this measure to still be functioning. For the HF chillers, EBCx had lowered the condenser water set point from 23.9 $^{\circ}$C (75 $^{\circ}$F) to 21.1 $^{\circ}$C (70 $^{\circ}$F), while raising the chilled water set point from 5.6 $^{\circ}$C (42 $^{\circ}$F) to 7.2 $^{\circ}$C (45 $^{\circ}$F). This measure was also found to be in operation at the time of this study.

For the JF buildings, air handling units and terminal boxes were scheduled at the time of EBCx to reflect occupancy patterns, scheduling unoccupied hours as 6 PM to 6 AM on weekdays and all day on weekends. At the time of this study, JF3 was evaluated, and the control was found to be working fairly well, with only a couple of override issues. Additional savings opportunities for the JF buildings were also identified in this study, including air flow and scheduling opportunities and control overrides that needed adjustment. For the HF chillers, the leaving condenser water set point was lowered as part of the EBCx project.

Overall at the Hawthorn Farms campus the EEMs were found to have been maintained, with the exception of the terminal unit reheat optimization in HF1. Of the original projected savings in the three buildings at Hawthorn Farms, 89 % of the electric savings and 0 % of the natural gas savings were still being achieved at the time of this study. In the five buildings at Jones Farm, the results were more mixed and less quantifiable. The recommended scheduling changes were still programmed at a high level, but it appeared that numerous control overrides at a zone or box level had been made. Some overrides may have been due to changes in space use (such as conversion to a lab), but in many instances conference and training rooms were maintaining occupied modes around the clock. The trending done on some of the variable speed air handlers showed little difference between day and nighttime airflow suggesting that terminal box scheduling was not having an impact on overall airflow.

Oregon: Summary

Of the eight EBCx projects studied in Oregon in 1999 and 2000 quantitative findings were reported for three buildings and qualitative findings for the other five. For the three buildings on the Hawthorn Farms campus, totaling 60 000 m^2 in floor area, 89 % of the original electric savings were achieved in 2004 and 0 % of the natural gas savings were achieved in 2004. For the five buildings on the Jones Farm campus with 130 000 m^2 of floor area, the results were mixed and less quantifiable. It was found that scheduling changes were still programmed at a high level, but numerous control overrides at a zone or box level had been made.

Colorado: Office Building in Colorado

A study completed in 2005 evaluated the persistence of EBCx savings in a large office building in Colorado (Selch 2005). Of the studies of this kind done to date, this study appears to have chosen the largest window of time over which to look at persistence. The office building was commissioned in 1995, which resulted in verified savings of 14 % in electrical demand, 25 % in electrical use, and 74 % in gas use. In 2003, the building was again commissioned, at which time the status of the energy conservation measures implemented in the initial commissioning effort was evaluated.

The computation of savings was done in two ways. The overall energy use of the building for each year was obtained from utility bills. These data were then normalized to account for factors such as weather differences, changing occupancy patterns in the building, and added construction in the building. In this way the yearly energy use could be accurately compared to the baseline, pre-commissioned energy use. The other savings calculation method was an individual measure evaluation. Specific measures that impacted individual HVAC system components were examined. To perform the calculations, Options B & C of the International Performance Measurement and Verification Protocol (IPMVP 2001) were employed, Option B being used for individual measure evaluation, and Option C for whole building usage comparison.

Table 3.9 summarizes the results of the individual measures evaluation. Using the same baseline, the savings achieved from the 1996 commissioning effort (20 %) are compared with savings achieved in 2003 (17 %). It was supposed that the persistence of savings was represented by the difference in achieved savings between the two commissioning efforts. Therefore, , the 2003 savings was calculated as a percentage of the original 1996 savings (i.e., (1,330,000/1,600,000)*100) to represent the persistence of savings (83 %).

As noted in Table 3.9, it was calculated that 83 % of the electrical use savings had persisted, while 86 % of the electrical demand savings had persisted. The results of the whole building energy use comparison appear in Figure 3.5 and Figure 3.6. The left chart in each figure represents the raw values, while the right chart displays adjusted, normalized values.

Table 3.9 Electric savings persistence summary

	Baseline use	1996 savings	2003 savings
Electricity	(80,000,000 kWh)	20 % (1,600,000 kWh)	83 % persistence (17 % savings) (1,330,000 kWh)
Demand	(1564 kW)	14 % (219 kW)	86 % persistence 12 % savings (188 kW)
Gas	-	74 %	Complete persistence

Figure 3.5 Annual electrical demand, raw and adjusted.

Figure 3.6 Annual electrical use, raw and adjusted.

The annual demand and consumption values that were adjusted to account for changing conditions indicated that the savings achieved from EBCx had largely persisted. This was concluded with greater confidence due to the corroboration of the independent measure analysis.

The study reported that a large majority of the energy savings measures implemented in the original commissioning effort had persisted, as had their resultant energy savings. This was in spite of changing conditions in the building, including a complete change in operation staff. It was concluded that ECMs of this nature can persist for at least eight years even with limited support from operators and staff. However, it was noted that continued, on-going support to the building staff as part of the original commissioning effort probably would have resulted in complete persistence of the savings achieved.

3.2.3. Persistence of Commissioning Measures in New Buildings

PECI PIER Study

In the summer of 2002, a study was completed that had begun in the fall of 2001 under a California Energy Commission Public Interest Energy Research (PIER) project (Friedman et al. 2003). The purpose of the study was to examine ten buildings that were commissioned at building start-up in order to address the persistence of benefits from the commissioning process. This study drew qualitative conclusions about the persistence of new building commissioning, focusing on three issues: how well the benefits of commissioning persisted, the reasons for declining performance, and the methods that can be used to improve the persistence of benefits achieved through commissioning. A quantitative assessment of persistence by measure ("this measure has an expected persistence of X years") was outside the scope of this project, since a large number of buildings would have been required to determine the figures for each measure.

To evaluate the persistence of commissioning benefits on new buildings, the buildings first had to be selected. To qualify for the study, the facility needed to have been commissioned as a new building or major retrofit between two and eight years prior to the study. Due to the difficulty in finding such buildings with adequate commissioning documentation in California, five buildings were selected in the Pacific Northwest, and five more in California. It was not feasible to limit the study to buildings that followed the full commissioning process, from pre-design through final acceptance and post-occupancy, as described in ASHRAE Guideline 1 (ASHRAE 1996). The most completely commissioned and documented buildings were sought, but these typically did not include design-phase commissioning.

For each building, three to eight items were identified that were documented to have been fixed during commissioning. The changes and repairs made during commissioning generally fell into three categories: hardware, control system, and documentation improvements. Due to the focus on energy savings measures in the study, the hardware and control system changes with the greatest energy implications were of highest interest, as well as measures dealing with comfort and reliability. The amount of documentation available for each measure was also a driving force in measure selection. It was necessary to only evaluate those measures that had actually been implemented and documented. Routine maintenance issues or measures deemed static once corrected (such as equipment disconnected from the power supply) were not looked at. With the limited amount of time and funding for the study, it was necessary to focus on measures whose current status could easily be compared to the as-commissioned status and which would affect energy consumption. Because of the bias in selecting these measures, and the underestimation of savings persistence due to the limited number of measures considered, the results of the study were presented qualitatively.

For purposes of the study, it was decided that if the measure resulted in better performance than the pre-commissioning condition, then the measure was said to have persisted, even if it had been adapted to meet real operating conditions of the building. In some cases the persistence of a measure was somewhat subjective.

The people with the most knowledge about the control system at each site were interviewed. Some sites were identified for site visits, and for the others a second interview was conducted to discuss the current status of the commissioning measures. Six of the buildings were visited, during which the persistence of the selected commissioning measures was investigated, and the work environment and resources available to the operations staff were evaluated.

PECI PIER - Results

It was found that the process of finding qualified buildings for the study in California was difficult. Qualified buildings were located more easily in Oregon, most likely because of the longer history of new building commissioning in the Pacific Northwest. California had numerous existing buildings involved in EBCx projects, but new buildings having undergone commissioning at least two years earlier were sparse. For many of the commissioned buildings considered for the study, commissioning reports had not been written, so the information that could have been used by operations personnel to more efficiently operate the building essentially was lost. Often times in lieu of a report, the commissioning activities would simply be placed on a "punch list" for maintenance personnel to work on, who, when they had completed them usually did not document the changes. In other buildings the reports had been written, but were not readily available to the operations staff, having been filed away in storage and not easily accessible. In many cases where documentation did exist, it was not clear when or if the commissioning measures had been implemented, as they were noted as "recommendations" or "pending." These issues led to the conclusion that the term "commissioning" had been applied to a variety of different activities, including troubleshooting items and checklists, indicating a lack of consistency in the way the term was being applied.

Table 3.10 summarizes the commissioning measures studied and their level of persistence. A light gray square indicates that the measure persisted, while a black square indicates that the measure did not persist. A square split in half horizontally indicates that more than one measure was investigated in the category.

Table 3.10 Persistence of equipment and controls fixed during commissioning.

BUILDING (year commissioned)	DOCUMENTS			CENTRAL PLANT				AIR HANDLING AND DISTRIBUTION								PREFUNCTIONAL TEST					OTHER		
	Commissioning report on site	Commissioning report used	Control sequences available	Chiller control	Cooling tower control	Boiler control	Hydronic control	Economizer control algorithm	Discharge air temperature reset	Simultaneous heating and cooling	VFD modulation	Dessicant cooling	Duct static pressure	Space temperature control	Terminal units	Piping and fitting problems	Valve modification	Wiring and instrumentation	Sensor placement or addition	Sensor error or failure	Scheduling	Skylight louver operation	Occupancy sensor
California Lab and Office 1 (1995)	no	-	yes																				
Office Building 1 (1996)	no	-	yes																				
Office Building 2 (1996)	no	-	no																				
Office Building 3 (1994)	yes	yes	no																				
Office Buidling 4 (1994)	no	-																					
Pacific Northwest Office Building 5 (1997)	no	-	yes																				
Medical Facility 1 (1998)	yes	yes	yes																				
Medical Facility 2 (1998)	yes	yes	yes																				
Lab and Office 2 (1997)	no	-	yes																				
Lab and Office 3 (2000)	no	-	no																				

Key to shading:

Black cell: Measure did not persist	
Grey cell: Measure persisted	
Split cells indicate multiple measures	

Across the ten buildings studied, patterns about the types of commissioning fixes that persisted emerged. For the 56 commissioning fixes selected, well over half of the measures persisted. It was not surprising that hardware fixes, such as moving a sensor or adding a valve, persisted. Furthermore, when control algorithm changes were reprogrammed, these fixes often persisted, especially when comfort was not compromised. Many design phase fixes may have persisted in a similar way, but these were not able to be studied since only one building was commissioned in the design phase.

The types of measures that tended *not* to persist were the control strategies that could easily be changed, such as occupancy schedules, reset schedules, and chiller staging. Four out of six occupancy schedules did not persist. Chiller control strategies did not persist in three out of four cases, most likely due to the complex nature of control in chilled water systems. The study of sensor issues was limited to major sensor problems that were corrected during commissioning, such as sensor failure or excessively faulty readings. With this selection bias applied, two out of five sensor repairs did not persist.

Among the commissioning measures implemented, a few cases involved technologies that were new or different from normal practice. Due to lack of documentation, these measures were not included in this study, but it was observed during the investigation that these measures generally did not persist. This was attributed to a lack of operator training for the technologies.

PECI PIER - Discussion

The study suggested three possible reasons for lack of persistence among some measures. The first was limited operator support and high operator turnover rates. Operators often did not receive the training necessary or they did not have sufficient time or guidance for assessing energy use, and the training given new operators who came in after the commissioning was usually inadequate. The second reason involved poor information transfer from the commissioning process. For nearly every case studied, the commissioning report was either difficult to locate, or was not even located on site, which reduced the ability of building operators to review commissioning measures implemented. The third reason for lack of persistence was a lack of systems to help track performance. Operators spent most of their time responding to complaints and troubleshooting problems, leaving little time to focus on assessing system efficiency. Aside from this, lack of information and knowledge impeded the efficiency assessment by building operators.

The persistence of commissioning benefits was found to be highly dependent on the working environment for building engineers and maintenance staff. A working environment that was supportive of persistence included adequate operator training, dedicated operations staff with the time to study and optimize building operation, and an administrative focus on building performance and energy costs. Trained operators were found to be knowledgeable about how the systems should run and, with adequate time and motivation to study the system operation, these operators evaluated and improved building performance. In five buildings, operators participated in the commissioning process and came away with a good understanding of their systems. In addition, good system documentation in the form of a system manual served as a troubleshooting resource for operators at two buildings. It was noted that administrative staff can help enable a supportive working environment by placing high priority on energy efficient systems and operator training. Only a few of the buildings studied seemed to operate in this environment, and the measures investigated at these facilities had the highest rate of persistence.

Some of the measures simply persisted by default – no maintenance being required to keep them operational. If comfort issues were not a factor, or the measure involved programming buried deep within code, the measures tended to persist.

The study recommended four methods for improving persistence. First, operators should be provided with training and support. Especially with high operator turnover, adequate training is needed for benefits to persist, and a working environment with energy efficiency as a high priority is also beneficial. Second, a complete systems manual should be provided at the end of the commissioning process. This will serve as a reference for building operators, and will allow the systems knowledge gained from the commissioning process to be available over the long term. Third, building performance should be tracked. New building commissioning efforts should help to implement mechanisms for performance tracking, including what information to track, how often to check it, and the magnitude of deviations to address. Fourth, commissioning should begin in the design phase to prevent nagging design problems. Changes made on paper before construction has begun tend to be more cost effective and have higher levels of persistence.

The study concluded with a recommendation that more in-depth, quantitative studies be performed to investigate the life of commissioning measures and carry out cost-benefit analyses for new building commissioning. It was further recommended that a manual of guidelines for improving persistence be developed to give guidance and direction to building operators with regard to energy efficiency.

3.2.4. Strategies for Improving Persistence in New and Existing Buildings

As a follow-up to the study of persistence of commissioning benefits for new buildings performed in California and Oregon, and the study of persistence of EBCx benefits done at Texas A&M (both described previously), a report was issued in July 2003 addressed to building owners, managers, and operators suggesting methods for improving the persistence of commissioning benefits for both new and existing buildings (Mills et al. 2004). The report began by summarizing the key conclusions of both studies, namely that many commissioning benefits tend to persist fairly well, but that significant opportunities still exist for improving overall savings persistence. The report then proposed that an emphasis on certain key elements of energy analysis and efficiency would pave the way for long-term success in building operation and energy use. In particular, seven recommendations were discussed at length: design review, building documentation, operator training, building benchmarking, energy use tracking, trend data analysis, and EBCx. A summary of the discussion of each of these topics is presented here.

Design Review

As much as one-third of major commissioning problems can be traced back to the design phase of the project, and these problems often plague building operators throughout the life of the building. Allowing professional engineers to review the design while it still in the design phase, is a cost-effective way to prevent future problems. Correcting design problems on paper is easier and less costly than attempting to correct them once the building is completed. Some of the issues to be considered in reviewing a design are test port location, equipment accessibility, load calculations and minimum flow settings, control system sequences and point lists, and standard design details. The process of design review should begin as soon as possible to allow opportunity for correction.

Building Documentation

Good system documentation is not a common practice currently in the construction environment. While it may seem like a costly and time-consuming effort, this documentation is the best way to ensure that the knowledge base obtained during design, construction, and commissioning of the building is preserved, and will aid in maintaining commissioning benefits. The three most vital items to document are the final design intent, the sequences of operation, and the system diagrams. Other important documents include the operator's log, commissioning summary report, general description of facility and systems, as-built documents, detailed description of each system, location of all control sensors and test ports, and capabilities and conventions of the DDC system. The best time for this documentation to occur is during the construction phase of the building. For existing buildings, a good time is during a retrofit or EBCx. The documentation should be compiled into a systems manual that is readily accessible.

Operator Training

Effective operator training will allow the benefits of building commissioning to persist, and will aid in preventing problems. Training opportunities exist for building operators during the commissioning process, through manufacturers and vendors, in operator certification programs, and using building documentation. It is also essential that new operators be trained sufficiently so that the knowledge gained by the previous operator is not lost. Some suggested training topics include: descriptions of equipment, equipment start-up and shut-down procedures, operation and adjustment of controls, review of system documentation, common troubleshooting problems, maintenance requirements and schedules, health and safety issues, special tools and spare parts inventory, and emergency procedures.

Building Benchmarking

Benchmarking refers to measuring the energy use of a building relative to other buildings, and provides a way to track energy use over time and compare it with the competition. This will allow building owners and operators to prioritize initiatives and improve energy efficiency. Several tools exist to aid in the benchmarking process. Two of these are the ENERGY STAR® Portfolio Manager, which uses a number of factors to make meaningful comparisons with other buildings under different conditions, and the Cal-Arch Building Energy Reference Tool, which is a quick and simple tool for comparing energy use per square foot.

Energy Use Tracking

Tracking utility bills or metered data is an effective method for recognizing energy use problems that may not result in comfort problems, and therefore might not be noticed any other way. It is essential for continued energy efficiency and persistence in commissioning benefits. The energy use curves should be compared for different years to look for patterns, anomalies, and peaks and valleys. An Energy Information System (EIS) is a useful tool for automating utility tracking. It saves time, provides immediate feedback, can gather additional data, and can allow access over the Internet.

Trend Data Analysis

DDC systems allow points to be trended over time. Knowing how to interpret these trended data is essential for identifying and correcting problems in building energy consumption and performance. The data should be examined regularly to determine if the system and its individual components are functioning as desired. Automated diagnostic tools exist to aid in this process, having automated capabilities in the following categories: data acquisition, archiving and pre-processing, detection, and diagnosis. Two tools available are ENFORMA and PACRAT. PACRAT can be used as an ongoing diagnostic tool.

Re-commissioning[14]

The process of re-commissioning, especially when it draws on building documentation and previous commissioning activities, is very effective in maintaining commissioning benefits. The time to consider re-commissioning largely depends on the effectiveness of operations and maintenance strategies and overall building performance. Commissioning can be performed by an outside commissioning provider when an outsider's view is considered helpful, or it may be done in house. In-house commissioning increases the knowledge level of those participating with regards to building operation. Continuous Commissioning® is an ongoing commissioning process developed by the Energy Systems Laboratory at Texas A&M University that has the same general goal as EBCx, but focuses strongly on the persistence of commissioning benefits.

The report concluded by reiterating the need to pursue the topics addressed during and after the commissioning process to maintain the benefits of commissioning over the long-term.

3.2.5. Related Reports

A report was compiled in 2004 that evaluated the cost effectiveness of commissioning in new and existing buildings (Mills et al. 2004). The largest study of its kind to date, it examined the results of commissioning for 224 buildings across 21 states. Among the existing buildings commissioned, a median payback period for commissioning was reported to be 0.7 years. For new buildings, this value was found to be 4.8 years. Both of these figures excluded non-energy benefits, which would increase the savings experienced.

While persistence of savings was not the primary focus of the study, it was examined briefly since it plays a role in determining overall savings. Figure 3.7 shows the persistence of savings results for 20 of the buildings in the study, with a four year period following commissioning in each building. The savings are indexed by a comparison of the year's consumption to the pre-commissioning baseline consumption. The savings are compared by category: electricity, fuel, chilled water, and steam/hot water.

---

[14] Re-commissioning is a term sometimes applied when EBCx is carried out on a building that was commissioned as part of the construction process. Some organizations use the term EBCx irrespective of whether or not the building was previously commissioned.

Figure 3.7 Emergence and persistence of energy savings (weather normalized).

An important factor noted in the report was the fact that in many cases of commissioning, the recommended measures were implemented gradually; indicating that the first year after commissioning was not the best year for calculating savings. On the other hand, it was also observed that after time some of the savings began to degrade due to changing building conditions, operations, or aging. As seen in the figure, the maximum value for savings was reached and subsequently savings began to degrade. This effect was smallest for electricity, but much more noticeable for chilled and hot water and steam.

With regard to persistence of commissioning benefits, the report concluded that tracking energy consumption for evidence of significant consumption increases is the most important means of determining the need for follow-up commissioning, and that while controls changes by building operators account for a portion of savings degradation, hidden component failures are perhaps the greatest culprit in persistence problems.

3.2.6. Methodologies for Determining Persistence of Commissioning Measures and Energy Benefits of Commissioning

The EBCx studies that provided a quantitative evaluation of the persistence of energy benefits of commissioning, used variations on two different approaches to evaluate the persistence of energy benefits.

The study of ten Texas buildings (Turner et al. 2001) used a variation on Option C of the IPMVP that normalized for weather differences between years by selecting a "normal" year of weather data in the sequence available that most closely met long-term norms. A suitable three-parameter or four-parameter regression model of the baseline year was created along with models of the performance of the building in each year evaluated.

Then the annual consumption for each year was determined by running the appropriate model with the appropriate year of weather data. The study of eight SMUD buildings (Bourassa 2004) used the same methodology, except that they used a long-term average weather year instead of selecting one of the available years of weather data. The Colorado study used a different approach, evaluating savings persistence with IPMVP Option C with baseline adjustments and IPMVP "Option B" was used to determine savings for specific measures in operation. The Oregon study did not specify how savings were evaluated.

The study of eight buildings in Oregon (Peterson 2005) and the Colorado building (Selch and Bradford 2005) used a different approach. These studies examined each of the measures that had been implemented and determined whether the measures were still in place and functioning. Peterson found that in three of the buildings, she could quantify the savings associated with measures that had been disabled after four years. It was found that numerous measures implemented in the other five buildings were still in place, but there were also numerous overrides and changes that had occurred as well. It was not possible to quantify the degree of persistence in these buildings. Selch and Bradford (2005) found that they were able to quantify the savings associated with measures that had been disabled.

The study of ten new buildings that had been commissioned in Oregon and Washington (Friedman et al. 2003b) used a methodology that quantified the number of measures that were still in place, but it did not seem appropriate to try to quantify the energy savings associated with these measures. The four EBCx studies all discussed the measures found to be still operating and those that had been changed. The Texas study used calibrated simulation to evaluate measures that had been changed. The other studies were not explicit in the methods used to evaluate the impact of measure changes.

3.2.7. Summary and Conclusions

The results of studies from five projects related to commissioning, either in new or existing buildings, described represent the extent of research that has been performed with regard to the persistence of commissioning benefits over time.

The savings in the buildings that were retro-commissioned generally showed some degradation with time. In retro-commissioned buildings, savings generally decreased with time, but there is wide variation from building to building. For the buildings where savings persistence was quantified:

- Savings persistence at the time of the study (3 years to 8 years after commissioning) ranged from about 50 % to 100 % in all but one or two buildings.
- Average savings at the time of the study were about 75 % of the original savings.
- The most dramatic savings degradation was caused by undetected mechanical or control component failures.

For the new buildings, well over half of the 56 commissioning fixes persisted. Hardware fixes, such as moving a sensor or adding a valve, and control algorithm changes that were reprogrammed generally persisted. Control strategies that could easily be changed, such as occupancy schedules, reset schedules, and chiller staging tended not to persist. It was also found that the extent to which persistence occurs is also related to operator training.

As is evident, the number of buildings studied in all of the papers described here represents a very small portion of commercial buildings that have undergone commissioning or EBCx. More research is needed to:

- Develop a uniform methodology for determining commissioning persistence.
- Determine the persistence of savings from a broader sample of buildings.
- Develop simple tools for tracking performance of commissioning measures.
- Develop practical methods for owners and operators to better maintain commissioning savings.

3.3 Influence of Savings Normalization Method on Persistence

3.3.1 Introduction

There are several different existing building commissioning processes including existing building commissioning (EBCx) and Continuous Commissioning® (CC®). The purpose of commissioning as detailed by ASHRAE is to ensure proper operation of a building according to the design intent (ASHRAE,1996). Existing building commissioning often involves making retrofits and optimizing building operation to achieve greater energy efficiency and energy savings.

The means used to determine these savings vary in both complexity and ease of performance. The International Performance Measurement and Verification Protocol (IPMVP) defines acceptable approaches for determining energy savings in buildings that have undergone energy conservation measures (ECMs) such as those carried out during commissioning (IPMVP 2002). Under the IPMVP there are four separate savings determination methods, Options A-D. Option A and Option B are not appropriate for determining whole building commissioning savings.

Option C of the IPMVP uses whole building data to develop consumption models such as regression models while Option D uses calibrated simulations to determine building energy consumption. Regression models are a quick way to relate heating and cooling consumption to the outside dry bulb temperature for a specific time period using that period's consumption data. Calibrated simulations require a series of inputs to a simulation tool that are adjusted, or calibrated, until the simulated consumption closely matches the heating and cooling data as a function of the outside dry bulb temperature. Calibrated simulations are valuable because they can identify and verify potential causes for changes in consumption from year to year but are much more time consuming than regression models.

Once an energy consumption model (regression model or calibrated simulation) that determines consumption as a function of outside dry bulb temperature has been obtained for baseline and post-commissioning periods, commissioning savings can be determined by two different weather normalization approaches:

- The standard International Performance Measurement and Verification Protocol (IPMVP) weather normalization approach (IPMVP 2002)
- The Normalized Annual Consumption (NAC) weather normalization approach (Fels 1986).

The standard IPMVP weather normalization approach calculates actual savings as the difference between the post-commissioning energy consumption determined by the pre-commissioning baseline model when using the post-commissioning weather data and the measured energy consumption during the post-commissioning period. In contrast, energy consumption models can be weather-normalized to "normal" or average weather conditions to mitigate the effects of varying weather from year to year. The common term for the annual consumption under the "normal" weather year is Normalized Annual Consumption or NAC (Fels 1986). Under the NAC weather normalization approach, each of the energy consumption models uses the "normal" weather year and savings are determined by the difference in the baseline and post-commissioning consumption.

Since the reported savings from commissioning are essential in telling the success of a commissioned building, it is important to know in some terms how the savings and persistence of savings results of one savings determination procedure compare to the other. In particular, it is useful to know whether use of the NAC weather normalization approach provides less variability in the persistence of commissioning savings than use of the standard IPMVP weather normalization approach when using a set of different weather years. Likewise, it is valuable to identify whether use of Option C of the IPMVP with regression modeling provides less variability in the persistence of commissioning savings than use of Option D of the IPMVP when using a set of different weather years. Persistence of savings refers to the degree to which post-commissioning savings are maintained from year-to-year. Specifically, the persistence of savings is the absence of change in savings between the first post-commissioning period and any later subsequent post-commissioning period.

Using the weather normalization approaches and IPMVP savings determination methods, about 30 existing buildings have been analyzed in previous studies to quantify savings persistence and identify reasons for changes in savings from year to year after commissioning has been performed. Similarly, over 100 buildings that have undergone major retrofits have been analyzed to quantify savings persistence.

Additional commissioned and retrofitted buildings have been previously analyzed where savings results are given without persistence results. Further research on additional buildings is important because a larger set of buildings documenting savings persistence can help identify ways to make commissioning savings persist longer and encourage more to take advantage of the benefits of this energy saving process.

The objectives of this chapter are to determine:
- Whether use of the NAC weather normalization approach provides less variability in commissioning savings and the persistence of commissioning savings than use of the standard IPMVP weather normalization approach, and to quantify any difference observed.
- Whether use of Option C of the IPMVP with regression modeling provides less variability in commissioning savings and the persistence of commissioning savings than use of Option D of the IPMVP, and to quantify any difference observed.

3.3.2 Savings normalization methods evaluated

Pre- and post-commissioning data were collected from the Civil Engineering/Texas Transportation Institute (CE/TTI) Building on the Texas A&M University campus. CE/TTI is used to compare the savings variability of the NAC and the standard IPMVP weather normalization approaches, and to compare the savings variability of Option C with regression modeling and Option D of the IPMVP. Only chilled and hot water results are used. Electricity results are not analyzed because electricity consumption is assumed to be mostly constant, independent of outside air temperature for the buildings on the Texas A&M University chilled and hot water campus loops.

Hourly chilled water and hot water data are used to calculate daily average consumption. For days in which 19 or more of the hours of data exist, the average from those hours is used as the daily average. This number is recommended by the Energy Systems Laboratory (ESL) data analysis group based on their experience. For days in which 18 or fewer hours of data exist, the daily average is determined by linear interpolation from daily averages of prior and subsequent days. Energy balance plots are used to screen the daily consumption data (Shao 2005). Note that this data screening method requires electricity, chilled water, and hot water data. Electricity data is used to screen for poor chilled water and hot water data but is not analyzed for reasons previously mentioned. Data identified as potentially erroneous are discarded. If one or more of the three data streams has poor data quality for a particular day then the other streams of data are discarded since the energy balance data screening method can only indicate good data quality for all or none of the three streams.

Review of the pre-commissioning data for CE/TTI shows very little accuracy or availability. In order to create an accurate pre-commissioning baseline, the commissioning report (Chen et al. 2004) is used to document key building and HVAC system parameters essential for creating a simulation for the pre-commissioning period by changing inputs of the calibrated simulation from the first post-commissioning period.

IPMVP Savings Methods

Once consumption data has been screened, Option C with regression models and Option D of the IPMVP can be used to determine savings. These savings methods use regression models and calibrated simulations to obtain a relationship for energy consumption as a function of the outside air dry bulb temperature. Three-parameter change-point (3P CP or PRISM) and four-parameter change-point (4P CP) regression models are used to model cooling and heating consumption. Using AirModel (Liu 1995), calibrated simulations are created that model chilled and hot water consumption.

Data for CE/TTI are separated into pre- and post-commissioning periods. Where possible, consumption data periods are divided into full calendar years. In some cases, however, periods are modified to be either shorter or longer than one calendar year where large periods of data are missing or of poor quality, or when the commissioning process takes place in the middle of a year. Katipamula et al. (1995) concluded that regression modeling of large commercial buildings can be accurate and reliable with at least three to six months of daily data. This data length requirement is met for all models and care is given to ensure that data spans a broad temperature range when less than a year of data is available. This process maximizes the amount of data that can be used for the study. As a result, each building has different pre- and post-commissioning period lengths. Consumption from each model is annualized by using a full weather year's temperature data, making comparison between periods of different lengths possible. A description of the weather data used in conjunction with the weather normalization approaches utilized follows the discussion of Options C and D.

Regression models are created for each of the post-commissioning periods using CE/TTI data. Calibrated simulations are also performed for each of CE/TTI's pre- and post-commissioning periods using AirModel. Since there are insufficient pre-commissioning data available to adequately model building chilled and hot water consumption, the first post-commissioning period consumption is simulated according to building and system characteristics stated in the commissioning report. After calibrating this simulation, the inputs are adjusted to reflect the pre-commissioning operation of the HVAC systems as documented in the commissioning report. In this manner, baseline consumption models are obtained for chilled and hot water. A regression model is also created from the calibrated simulation baseline output for both chilled and hot water to determine savings with Option C of the IPMVP.

Option C with Regression Models

The regression models used under Option C of the IPMVP to determine consumption in this study use four-parameter change point models. These models find energy consumption expressions as a function of daily average temperature. Chilled and hot water typically employ a three- or four-parameter change point model in commercial buildings.

Equation 1 expresses the functional form of four parameter models.

$$E = a + b_1 * (T_{OA} - T_{CP})^- + b_2 * (T_{OA} - T_{CP})^+ \qquad (1)$$

where a is the energy use at the change point temperature, T_{cp}, and b is the slope. T_{OA} is the ambient temperature. The notation $( )^{+/-}$ indicates that the quantities within the parenthesis should be positive or negative as the sign indicates; otherwise they are set to zero.

Option D

Option D of the IPMVP uses calibrated simulations to determine energy savings. When using AirModel to simulate building energy consumption, the user must specify two files, the input file and the weather source file. The input file includes specific quantities for the building and system parameters and characteristics such as conditioned floor area, room temperature, cold deck temperature, total and outside air flow settings, and night-time setback schedules. For this study, the weather file includes daily averaged values for dry bulb temperature and dew point temperature, although AirModel has the option of entering hourly values.

After running the AirModel simulation, the simulated output must be calibrated to the measured consumption data. The basis of the simulation inputs for CE/TTI comes mainly from commissioning reports, building blue prints, and trips to the building for assessment. A brief description of the calibration process is given here. A more detailed procedure of the simulation calibration process is given by Claridge et al. (2003) and Wei et al. (1998).

The term "calibration signature" (Claridge et al. 2003) is defined as follows:

$$CalibrationSignature = \frac{-residual}{MaxiumumMeasuredEnergy} \times 100\%$$ (2)

where

$$residual = SimulatedConsumption - MeasuredConsumption$$ (3)

The maximum measured energy is the maximum heating or cooling energy use recorded over the temperature range of the particular data file being used. The calibration signature is a normalized plot of the difference between measured energy use and simulated energy use over a specified temperature range. For each temperature, a measured energy use value and a simulated energy value exist. The difference in these values for each point is divided by the maximum measured energy use and multiplied by 100 %. These values are then plotted versus temperature.

The calibration signature is now compared to published characteristic signatures of the given HVAC system type in the given climate. A characteristic signature is identical to a calibration signature, except that instead of comparing simulated and measured values, it compares two simulations. One simulation is taken to be the baseline or "measured" value. Then, by varying parameters one by one, signatures can be plotted and compared. Characteristic signatures (Claridge et al. 2003) are defined as:

$$CharacteristicSignature = \frac{ChangeInEnergyConsumption}{MaximumEnergyConsumption} \times 100\%$$ (4)

As mentioned, the baseline model is treated as the "measured" case, and maximum energy consumption comes from this model.

Characteristic signatures can be generated for each HVAC system type. The majority of the CE/TTI building's HVAC systems are single-duct variable-air-volume (SDVAV), thus making it most practical to refer to SDVAV characteristic signatures when calibrating simulations. The parameters of major importance for which characteristic signatures should be generated include minimum air flow rate (VAV systems), floor area, preheat temperature, internal gains, outside air flow rate, room temperature, envelope U-value, and economizer.

Two indices used for evaluating the accuracy of a simulation are the "Root Mean Square Error" and the "Mean Bias Error." The Root Mean Square Error (RMSE) is defined as:

$$RMSE = \sqrt{\frac{\sum_{i=1}^{n} residual_i^2}{n-2}}$$ (5)

where n is the number of total data points. The RMSE is a good measure of the overall magnitude of the errors, but does not give any reflection of bias, since no indication is made as to whether the errors are positive or negative. A good simulation minimizes the RMSE and can achieve 10 % to 20 % CV-RMSE (IPMVP 2002). It is generally difficult to reduce this to smaller than 5 % to10 % CV-RMSE (Claridge et al. 2003). The Mean Bias Error (MBE) is defined as:

$$MBE = \frac{\sum_{i=1}^{n} residual_i}{n}$$ (6)

where n is the number of data points. The MBE is an overall measure of how biased the data is, since positive and negative errors cancel each other out. The MBE should be minimized in calibrating a simulation and should be less than ±20 % of the mean consumption (IPMVP 2002).

Calibration signatures combined with characteristic signatures are used to quickly calibrate a simulation. The calibration signatures for heating and cooling generated for the simulation are compared with the characteristic signatures from the corresponding system and climate type, to see which change of parameter or parameters most closely resembles the calibration signature. Normally one parameter is changed at a time in the correct direction and according to the magnitude needed. For example, if the calibration signature is in the range of 20 % for low temperatures, and a similar characteristic signature shows the same trend, but is in the range of only 5 %, the parameter adjustment would need to be significantly greater than what was done to get the characteristic signature in order to increase the magnitude. The adjustment is of course limited by reasonable values.. Once the parameter has been decided on, it is changed and the simulation is run again. The RMSE is calculated again, and calibration signatures are again generated and compared with the characteristic signatures. This process is repeated until the RMSE is minimized, and the calibration signature is flat and settled around zero. At this point, the simulation can be considered to be calibrated to the measured data. In most cases, however, it is difficult to obtain a completely flat calibration signature for both cooling and heating consumption. A well calibrated simulation has a CV-RMSE of 10 % to 20 % (Claridge et al. 2003) and the MBE should be less than 20 % of the mean consumption (IPMVP 2002).

Weather Normalization Approaches

The energy consumption models are now used to determine savings with two weather normalization approaches—NAC and standard IPMVP. In order to obtain a measure of variability in commissioning savings between the two weather normalization approaches, a set of different weather years is obtained to drive the energy consumption models. A set of different savings results is then obtained with both the NAC and standard IPMVP weather normalization approaches.

The weather year's data representing College Station, TX and used to create a set of savings results for the NAC and standard IPMVP weather normalization approaches were obtained from a few sources, depending on the year. For years 1973 to 2005, hourly weather data from the Easterwood Airport weather station was retrieved from the National Climatic Data Center (NCDC) weather database. For the years 1997 through June of 2004, NCDC data is used in conjunction with weather data from the Energy Systems Laboratory (ESL) database to fill in any missing data points.

As with the energy consumption data, hourly dry bulb and dew point temperature data are used to calculate daily average values. For days in which 19 or more of the hours of data exist, the average from those hours is used as the daily average. For days in which 18 or fewer hours of data exist, the daily average is determined by linear interpolation from daily averages of prior and succeeding days.

Additionally, daily average weather data from all available weather years are averaged to form a long-term average weather year. Each day's data of this long-term average weather year represents the average of all weather years' data for that day. For example, the daily average data for January 21 of the long-term average weather year is the average of all January 21 data from the existing weather years' data.

NAC Weather Normalization Approach

The NAC weather normalization approach determines savings as the difference between pre- and post-commissioning model consumption during a "normal" weather year. By using the same weather across all pre- and post-commissioning models, the variation in the consumption due to different weather patterns from year to year is minimized. Generally, long-term average weather data is used as the "normal" weather year when using the NAC weather normalization approach. This study, however, uses each of 29 different College Station, Texas weather years obtained from NCDC as the "normal" weather year. Each of these 29 weather years is used with every one of CE/TTI's pre- and post-commissioning energy consumption models (both regression models and calibrated simulations) to obtain 29 sets of normalized annual consumption. The savings are then determined in each of the post-commissioning periods for each weather year used.

Standard IPMVP Weather Normalization Approach

The standard IPMVP weather normalization approach employed to determine actual savings from commissioning activities calculates the difference between post-commissioning energy consumption determined by the baseline model with the post-commissioning period weather and the measured energy consumption taken from the post-commissioning period. In order to annualize the measured energy consumption, the model created from consumption data is used to determine the annual measured energy consumption by using the full weather year's ambient temperature data to drive the model. A more typical procedure in cases where there is missing data in a post-commissioning time period is to use the post-commissioning model to generate any missing data to add to the actual measured data. This approach is not used in this study, however, due to the necessity of using post-commissioning time periods of less or greater than one year in several instances.

Since the standard IPMVP weather normalization approach uses the measured post-commissioning energy consumption to determine savings, there is just one set of savings for each post-commissioning period. In order to form a larger sample size of savings results from the standard IPMVP weather normalization approach to compare to the NAC weather normalization approach, a method is employed that randomly selects a College Station weather year from the NCDC weather years retrieved as the 1st post-commissioning year, another as the 2nd post-commissioning year, yet another as the 3rd post-commissioning year, and so forth. As an example of this methodology, assume that a random run of weather years selected to find savings for the six CE/TTI post-commissioning periods are 1984, 1976, 1999, 1998, 1993, and 1974. The 1st post-commissioning period savings under the standard IPMVP weather normalization approach would be determined by subtracting the "measured" 1st post-commissioning period consumption determined by the period's model normalized to 1984 weather from the consumption of the baseline model normalized to 1984 weather. The 2nd post-commissioning period savings under the standard IPMVP weather normalization approach would be determined by subtracting the "measured" 2nd post-commissioning period consumption determined by the period's model normalized to 1976 weather from the consumption of the baseline model normalized to 1976 weather. The 1999, 1998, 1993, and 1974 weather years would similarly be used to determine the savings of the 3rd, 4th, 5th, and 6th post-commissioning periods, respectively. Other sets of random runs of weather years are used to obtain a set of savings results to determine whether the NAC weather normalization approach provides less variability in the persistence of commissioning savings than the standard IPMVP weather normalization approach. The 29 specific sets of random runs are given in Table 3.11. It should be reemphasized that the baseline regression model for CE/TTI used here is created from the synthetic "data" of the baseline calibrated simulation output.

Table 3.11 Sequence of College Station weather years for 29 different random runs used with both Option D and Option C with regression models in conjunction with the standard IPMVP weather normalization approach.

Run	1997	1998	1/1999-4/2000	2001	1/2002-11/2002	9/2003-6/2004
1	1993	1993	1980	1985	1984	2001
2	1984	1976	1999	1998	1993	1974
3	1977	1998	2001	1984	1991	1999
4	1979	2005	1978	2005	1978	1991
5	1992	1981	2003	2005	1998	1998
6	1980	1996	1975	1979	1985	2004
7	1976	2000	1992	1981	1990	2000
8	1990	1973	1999	2004	2002	Avg Yr
9	Avg Yr	Avg Yr	1996	2002	1985	Avg Yr
10	1990	1983	1997	1999	1978	1990
11	2000	1997	1994	1977	1983	1999
12	2004	1993	1996	1973	1984	2001
13	1991	1985	1990	2005	1994	2003
14	1991	2000	2000	1984	1985	2001
15	1973	1998	1990	1999	1997	1997
16	1993	1975	2003	2002	1975	2004
17	1974	1978	2001	2000	1985	Avg Yr
18	1975	1981	1973	1997	2004	1984
19	1999	Avg Yr	1996	1999	2002	2000
20	2004	1973	2004	1998	2002	1990
21	Avg Yr	1985	1979	1985	1996	1984
22	1992	1985	2002	2001	1989	1992
23	1994	1989	1989	1976	1991	1981
24	1981	Avg Yr	1999	2001	1978	2001
25	2004	1990	1977	1999	2000	1979
26	1979	1996	1980	1998	1977	1978
27	2004	1993	1979	2003	1978	2000
28	1994	2000	1997	1981	1999	2003
29	1999	1983	1973	1996	1975	1996

CE/TTI Regression Models and Calibrated Simulations

The CE/TTI Building has been commissioned twice. The first commissioning took place between August 1996 and September 1996. The second commissioning took place between December 2002 and August 2003.

Post-commissioning regression models and calibrated simulations are created for time periods where consumption data is available. Post-commissioning time periods are divided into full calendar years when possible. A period of missing hot water data, however, as well as the nine-month second commissioning period make this difficult to follow and consumption data period lengths are altered to lengths both shorter and longer than 12 months. The following is a list of the post-commissioning time periods for which regression models and calibrated simulations are created:

1. 1997
2. 1998
3. January 1, 1999 toApril 24, 2000
4. April 24, 2001to December 31, 2001
5. January 1, 2002 to November 30, 2002
6. September 24, 2003 to June 22, 2004

Table 3.12 summarizes the goodness-of-fit measures for each time period's calibrated simulation. The 1996 pre-commissioning simulation has no goodness-of-fit measures because it has no measured consumption data to be compared to.

Table 3.12 Goodness-of-fit measures for CE/TTI AirModel calibrated simulations.

RMSE	1996 Pre-Cx	1997	1998	1/1999-4/2000	4/2001-12/2001	1/2002-11/2002	9/2003-6/2004
CHW (MMBtu/day)	n/a	4.4687	4.6170	4.2340	4.6359	4.4325	3.9846
CHW (GJ/day)	n/a	4.7147	4.8712	4.4671	4.8912	4.6766	4.2040
HW (MMBtu/day)	n/a	2.0499	2.3970	2.3941	2.0091	2.7280	4.0636
HW (GJ/day)	n/a	2.1628	2.5290	2.5260	2.0097	2.8782	4.2873
MBE	**1996 Pre-Cx**	**1997**	**1998**	**1/1999-4/2000**	**4/2001-12/2001**	**1/2002-11/2002**	**9/2003-6/2004**
CHW (MMBtu/day)	n/a	-0.0381	-0.1636	0.4876	-0.0088	-0.6075	-0.1620
CHW (GJ/day)	n/a	-0.0402	-0.0173	0.5145	0.0093	-0.6410	-0.1702
HW (MMBtu/day)	n/a	-0.2631	-0.4496	-0.0782	0.5751	0.2108	0.9088
HW (GJ/day)	n/a	-0.2776	-0.4743	-0.0825	0.6068	0.2224	0.9588
CV-MBE	**1996 Pre-Cx**	**1997**	**1998**	**1/1999-4/2000**	**4/2001-12/2001**	**1/2002-11/2002**	**9/2003-6/2004**
CHW	n/a	-0.09 %	-0.35 %	1.23 %	-0.02 %	-1.32 %	-0.49 %
HW	n/a	-2.96 %	-6.13 %	-1.05 %	8.45 %	2.12 %	9.52 %
CV-RMSE	**1996 Pre-Cx**	**1997**	**1998**	**1/1999-4/2000**	**4/2001-12/2001**	**1/2002-11/2002**	**9/2003-6/2004**
CHW	n/a	10.04 %	9.78 %	10.67 %	9.65 %	9.66 %	12.10 %
HW	n/a	23.03 %	32.66 %	32.02 %	29.51 %	27.42 %	42.56 %

All of the baseline and post-commissioning chilled and hot water models are four parameter change point models (4P CP). Regression model goodness-of-fit measures are found in Table 3.13.

Table 3.13 Goodness-of-fit measures for CE/TTI regression models.

CHW	1996 Pre-Cx	1997	1998	1/1999-4/2000	2001	1/2002-11/2002	9/2003-6/2004
MBE (MMBtu/day)	n/a	-0.0002	-0.0002	0.0000	0.0938	0.0001	-0.0692
MBE (GJ/day)	n/a	-0.0002	-0.0002	0.0000	0.0889	0.0000	-0.0656
RMSE (MMBtu/day)	n/a	4.9755	5.2263	5.0009	4.7574	4.7104	4.1297
RMSE (GJ/day)	n/a	4.7158	4.9536	4.7399	4.5091	4.4646	3.9141
CV-RMSE	n/a	11.17 %	11.07 %	12.61 %	9.90 %	10.27 %	12.54 %
HW	**1996 Pre-Cx**	**1997**	**1998**	**1/1999-4/2000**	**2001**	**1/2002-11/2002**	**9/2003-6/2004**
MBE (MMBtu/day)	n/a	0.0005	-0.0003	0.0000	0.0063	0.0000	-0.2100
MBE (GJ/day)	n/a	0.0005	-0.0003	0.0000	0.0060	0.0000	-0.1990
RMSE (MMBtu/day)	n/a	1.7554	2.4857	2.0631	1.5637	2.2109	3.4344
RMSE (GJ/day)	n/a	1.6638	2.3560	1.9554	1.4821	2.0955	3.2552
CV-RMSE	n/a	19.72 %	33.87 %	27.60 %	22.97 %	22.22 %	35.97 %

It is interesting to note that the goodness-of-fit measures results of the calibrated simulations and regression models in Table 3.12 and Table 3.13 show that the calibrated simulations generally have a smaller RMSE and CV-RMSE than the regression models for chilled water. The results for hot water, however, show the opposite occurs—regression models generally have lower RMSE and CV-RMSE values than calibrated simulations. The significance of this result, however, is difficult to ascertain because AirModel links the chilled water and hot water consumption together while chilled water and hot water regression models are created independent of each other.

Results

The chilled water and hot water consumption, savings, percent savings, and change in percent savings are listed in Table 3.14. These values are determined with Option C of the IPMVP using the regression models and the NAC weather normalization approach. The long-term average College Station weather year is used as the "normal" weather year.

Table 3.14 CE/TTI chilled water and hot water consumption, savings, percent savings, and change in percent savings using the NAC weather normalization approach and Option C with regression models.

Year/Period	1996 Pre-Cx	1997	1998	1/1999-4/2000	2001	1/2002-11/2002	9/2003-6/2004
Values in MMBtu/yr							
CHW Use	17356	16491	15849	14890	14767	15822	13256
CHW Savings	Baseline	864	1507	2466	2589	1534	4100
CHW % Savings	Baseline	5.0 %	8.7 %	14.2 %	14.9 %	8.8 %	23.6 %
CHW Change in % Savings	n/a	n/a	3.7 %	5.5 %	0.7 %	-6.1 %	14.8 %
HW Use	3625	2804	2770	2553	3127	3828	2923
HW Savings	Baseline	821	856	1072	498	-203	702
HW % Savings	Baseline	22.7 %	23.6 %	29.6 %	13.7 %	-5.6 %	19.4 %
HW Change in % Savings	n/a	n/a	0.9 %	6.0 %	-15.9 %	-19.3 %	24.9 %
Values in GJ/yr							
CHW Use	16450	15630	15022	14113	13996	14996	12564
CHW Savings	Baseline	819	1428	2337	2453	1454	3886
CHW % Savings	Baseline	5.0 %	8.7 %	14.2 %	14.9 %	8.8 %	23.6 %
CHW Change in % Savings	n/a	n/a	3.7 %	5.5 %	0.7 %	-6.1 %	14.8 %
HW Use	3436	2658	2625	2420	2964	3628	2770
HW Savings	Baseline	778	811	1016	472	-192	665

Table 3.14 shows favorable savings results for chilled water and hot water. The chilled water and hot water each experience savings increases from the first to second and from the second to third post-commissioning periods. Overall, aggregate site savings decline sharply between the 2001 and 1/02-11/02 periods, dropping from 14.6 % to 7.1 %. Hot water savings show an especially sharp drop between these two periods, dropping from 13.7 % to -5.6 %. The second commissioning of the building, performed after the 1/02-11/02 period, appears to be worthwhile, as the hot water and aggregate site savings increase to 19.4 % and 15.6 %, respectively, based on the 1996 pre-commissioning baseline. While only the first year of post-commissioning data is available for the second building commissioning, the aggregate site savings return to a level similar to the peak achieved before the 2nd commissioning occurred. Over six post-commissioning periods, CE/TTI averages 12.5 % chilled water savings, 17.2 % hot water savings, 7.9 % electricity savings, and 11.4 % aggregate site savings.

SAVINGS AND VARIABILITY OF SAVINGS FROM DIFFERENT SAVINGS METHODOLOGIES RESULTS

Option D MBE Adjustment

Each of the calibrated simulations has an associated mean bias error (MBE) that if left unadjusted may significantly affect the post-commissioning savings depending on the magnitude of the MBE and its sign (positive or negative). The regression models created have essentially no MBE and consequently are not adjusted. In order to avoid biased savings and persistence results, adjustments are made to each of the annual consumption values determined by the calibrated simulations to offset the MBE of the calibrated simulation. Table 3.15 shows the MBE of the calibrated simulations from the pre-commissioning period and each of the post-commissioning periods. The annual adjustment given to each period's consumption determined with the calibrated simulations is also shown. The annual adjustment represents the opposite (positive or negative) of the MBE expressed as a daily value multiplied by 365 (days/yr).

Table 3.15 Calibrated simulation MBE and corresponding annual consumption adjustment for chilled and hot water.

Year/Period	1996 Pre-Cx	1997	1998	1/1999-4/2000	2001	1/2002-11/2002	9/2003-6/2004
Values in MMBtu							
CHW MBE (MMBtu/day)	-0.0381	-0.0381	-0.1636	0.4876	-0.0088	-0.6075	-0.1620
Annual CHW Adjustment (MMBtu/yr)	13.90	13.90	59.70	-177.96	3.23	221.75	59.14
HW MBE (MMBtu/day)	-0.2631	-0.2631	-0.4496	-0.0782	0.5751	0.2108	0.9088
Annual HW Adjustment (MMBtu/yr)	96.031	96.031	164.1	28.53543	-209.925	-76.9577	-331.702
Values in GJ							
CHW MBE (GJ/day)	-0.0361	-0.0361	-0.1551	0.4622	-0.0083	-0.5758	-0.1535
Annual CHW Adjustment (GJ/yr)	13.17	13.17	56.58	-168.67	3.06	210.18	56.05
HW MBE (GJ/day)	-0.2494	-0.2494	-0.4261	-0.0742	0.5451	0.1998	0.8614
Annual HW Adjustment (GJ/yr)	91.019	91.019	155.5	27.04626	-198.970	-72.9415	-314.392

NAC Versus Standard IPMVP Weather Normalization Approach

The percent savings results from the NAC weather normalization approach generally show good agreement with the percent savings results from the standard IPMVP weather normalization approach. Chilled water consumption and savings results are shown for both the NAC and standard IPMVP weather normalization approaches in Table 3.16 with Option C using regression models and Table 3.17 with Option D (MBE adjusted). The NAC weather normalization approach in these tables uses the long-term average College Station weather year. The differences in percent savings between the two weather normalization approaches shown in the tables for each post-commissioning period vary but there are only two post-commissioning periods where the difference is greater than 1 % and one post-commissioning period where the difference is greater than 2 %. However, some of these differences are relatively large compared to the percent savings of these post-commissioning periods. For example, the average 1997 percent savings between the two weather normalization approaches using Option D (MBE adjusted) are 4.34 % (see Table 3.16). The difference in savings, 0.58 %, represents 13.4 % of the average savings. In other words, while percent savings differences shown in Table 3.16 and Table 3.17 may seem small, a small difference may be significant if the overall savings is not that large. The chilled water percent savings differences between the NAC and standard IPMVP weather normalization approaches using Option D (MBE adjusted) average 0.78 % over all post-commissioning periods while the differences using Option C with regression models average 0.64 %.

Table 3.16 Chilled water consumption and savings results using Option D (MBE adjusted) of IPMVP to compare NAC and standard IPMVP weather normalization approaches. The long-term average weather year is used for the NAC weather normalization approach.

	Year/Period	1996 Pre-Cx	1997	1998	1/1999- 4/2000	2001	1/2002- 11/2002	9/2003- 6/2004
NAC Weather Normalization Approach								
Normalized to Long-Term Avg Weather	CHW Use (MMBtu/yr)	16800	16024	15409	14745	14469	15639	13069
	CHW Use (GJ/yr)	15923	15188	14605	13976	13714	14823	12387
	CHW Savings (MMBtu/yr)	Baseline	777	1392	2056	2331	1161	3732
	CHW Savings (GJ/yr)	Baseline	736	1319	1949	2209	1100	3537
	CHW % Savings	Baseline	4.62 %	8.28 %	12.24 %	13.87 %	6.91 %	22.21 %
Standard IPMVP Weather Normalization Approach								
Baseline Consumption with Post-Cx Weather	CHW Use (MMBtu/yr)	18874	17017	18610	17571	17839	17685	17983
	CHW Use (GJ/yr)	17889	16129	17639	16654	16908	16762	17045
Post-Cx Consumption with Own Period's Weather	CHW Use (MMBtu/yr)	18874	16253	17066	15685	15305	16164	13819
	CHW Use (GJ/yr)	17889	15405	16175	14866	14506	15320	13098
	CHW Savings (MMBtu/yr)	Baseline	764	1544	1886	2534	1522	4164
	CHW Savings (GJ/yr)	Baseline	724	1463	1788	2402	1443	3947
	CHW % Savings	Baseline	4.05 %	8.18 %	9.99 %	13.43 %	8.06 %	22.06 %
	CHW % Savings Difference	Baseline	0.58 %	0.10 %	2.24 %	0.45 %	1.15 %	0.15 %

Table 3.17 Chilled water consumption and savings results using Option C of IPMVP with regression models to compare NAC and standard IPMVP weather normalization approaches. The long-term average weather year is used for the NAC weather normalization approach.

	Year/Period	1996 Pre-Cx	1997	1998	1/1999-4/2000	2001	1/2002-11/2002	9/2003-6/2004
NAC Weather Normalization Approach								
Normalized to Long-Term Average Weather	CHW Use (MMBtu/yr)	17356	16491	15849	14890	14767	15822	13256
	CHW Use (GJ/yr)	16450	15630	15022	14113	13996	14996	12564
	CHW Savings (MMBtu/yr)	Base-line	864	1507	2466	2589	1534	4100
	CHW Savings (GJ/yr)	Base-line	819	1428	2337	2454	1454	3886
	CHW % Savings	Base-line	4.98 %	8.68 %	14.21 %	14.92 %	8.84 %	23.62 %
Standard IPMVP Weather Normalization Approach								
Baseline Consumption with Post-Cx. Weather	CHW Use (MMBtu/yr)	18860	17142	18840	18034	18058	17841	18094
	CHW Use (GJ/yr)	17876	16247	17857	17093	17116	16910	17150
Post-Cx Consumption with Own Period's Weather	CHW Use (MMBtu/yr)	18860	16253	17059	15685	15298	16125	13734
	CHW Use (GJ/yr)	17876	15405	16169	14866	14500	15283	13017
	CHW Savings (MMBtu/yr)	Base-line	889	1780	2349	2760	1716	4360
	CHW Savings (GJ/yr)	Base-line	843	1687	2226	2616	1626	4132
	CHW % Savings	Base-line	4.71 %	9.44 %	12.45 %	14.64 %	9.10 %	23.12 %
	CHW % Savings Difference	Base-line	0.27 %	0.76 %	1.76 %	0.28 %	0.26 %	0.50 %

While the results in Tables 3.16 and 3.17 do not show any striking differences in savings between the NAC and standard IPMVP weather normalization approaches, they do not show which approach has less variability. Figure 3.8 compares chilled water plus hot water percent savings quartiles of the NAC and standard IPMVP weather normalization approaches side by side when using the 29 different College Station weather years and random runs. Option D (MBE adjusted) of the IPMVP is used for both weather normalization approaches in Fig. 3.9. The minimum, median and maximum values are shown. Figure 3.8 is significant in that it shows an appreciably smaller variability in savings for the NAC weather normalization approach than the standard IPMVP approach for several of the post-commissioning time periods. It shows that depending on the weather years used for the standard IPMVP weather normalization approach, there may be considerably less persistence in savings over time than there would be if the NAC weather normalization approach is used.

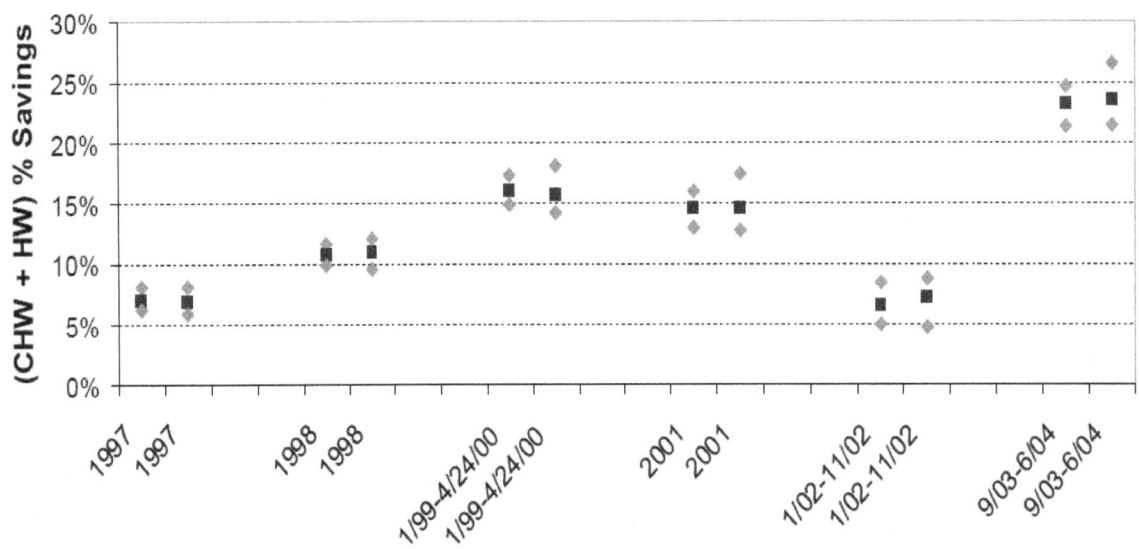

Figure 3.8 Post Commissioning Period, chilled water plus hot water percent savings variability (min, median and max) comparison between NAC (left set) and standard IPMVP (right set) weather normalization approaches when using Option C of IPMVP with regression.

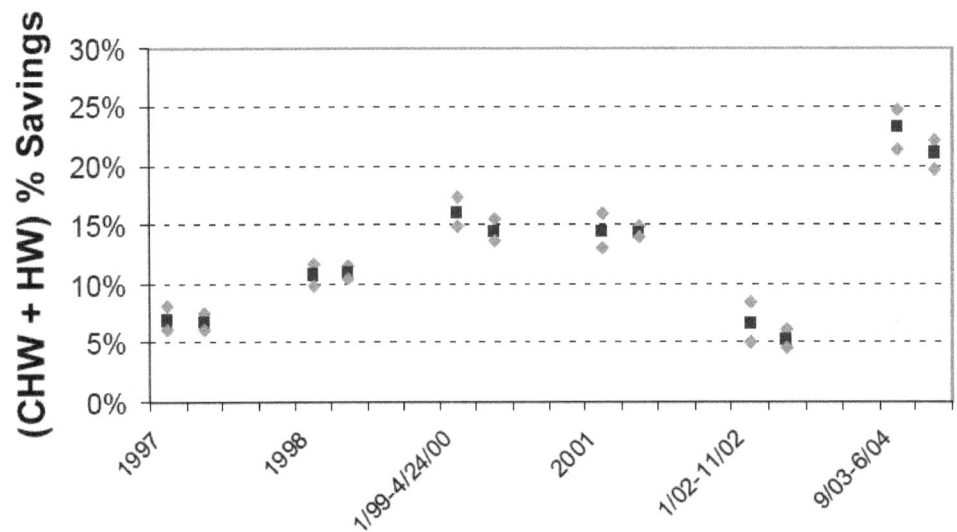

Figure 3.9 Post Commissioning Period, chilled water plus hot water percent savings variability comparison between NAC (left set) and standard IPMVP (right set) weather normalization approaches when using Option D (MBE adjusted) of IPMVP.

Chilled and hot water savings ranges and averages across all weather years with the NAC weather normalization approach and all random runs with the standard IPMVP weather normalization approach are listed in Table 3.18. Option D (MBE adjusted) of the IPMVP is used for both weather normalization approaches in this table. The results show that the NAC weather normalization approach has a smaller average range in savings across all post-commissioning periods than the standard IPMVP weather normalization approach for both chilled and hot water. For chilled water, the mean percent savings range is 1.32 % for the NAC weather normalization approach and 2.64 % for the standard IPMVP weather normalization approach. For hot water, the mean percent savings range is 3.30 % for NAC and 5.04 % for standard IPMVP. Despite these differences, the mean of the post-commissioning period average savings is quite similar. For chilled water, the mean of the average percent savings is 11.38 % for the NAC weather normalization approach versus 11.40 % for the standard IPMVP weather normalization approach. For hot water, the mean of the average percent savings is 15.99 % for both the NAC and standard IPMVP weather normalization approaches. The differences between the two weather normalization approaches in average percent savings across each of the post-commissioning periods are also relatively small. For chilled water, the largest difference in average savings between the two weather normalization approaches in a post-commissioning period is 0.25 %, occurring in the 9/03-6/04 period. For hot water, it is 0.28 %, occurring in 2001.

Table 3.18 Chilled and hot water percent savings range and average across all College Station weather years under NAC weather normalization approach and across all random runs under standard IPMVP weather normalization approach. Both approaches use Option D (MBE adjusted)

CHW % Savings Range	1997	1998	1/99-4/24/00	2001	1/02-11/02	9/03-6/04	Average
NAC	1.30 %	0.81 %	1.49 %	1.04 %	1.42 %	1.84 %	1.32 %
Standard IPMVP	0.82 %	1.67 %	2.62 %	3.47 %	1.64 %	5.62 %	2.64 %

HW % Savings Range	1997	1998	1/99-4/24/00	2001	1/02-11/02	9/03-6/04	Average
NAC	4.43 %	3.53 %	2.76 %	2.70 %	3.29 %	3.09 %	3.30 %
Standard IPMVP	4.31 %	5.02 %	9.00 %	6.11 %	2.77 %	3.03 %	5.04 %

CHW % Savings Average	1997	1998	1/99-4/24/00	2001	1/02-11/02	9/03-6/04	Average
NAC	4.33 %	8.21 %	12.12 %	14.02 %	7.18 %	22.41 %	11.38 %
Standard IPMVP	4.32 %	8.14 %	12.03 %	14.12 %	7.16 %	22.66 %	11.40 %

HW % Savings Average	1997	1998	1/99-4/24/00	2001	1/02-11/02	9/03-6/04	Average
NAC	18.71 %	23.50 %	25.36 %	16.04 %	-3.20 %	15.53 %	15.99 %
Standard IPMVP	18.87 %	23.77 %	25.19 %	15.76 %	-3.18 %	15.52 %	15.99 %

Option C with Regression Models Versus Option D of IPMVP

The previous section compares the variability in savings persistence between the NAC and standard IPMVP weather normalization approaches. It is also valuable to know whether Option C with regression models or Option D of the IPMVP shows less variability in the persistence of savings and by how much. Knowing this may influence how much time one is willing to invest calibrating simulations under Option D when one could quickly create regression models.

Table 3.19 quantifies the variation in percent savings across the different weather years using the NAC weather normalization approach for both Option C with regression models and Option D (MBE adjusted) of the IPMVP. The range and average percent savings across all 29 College Station weather years for each post-commissioning period are shown. Table 3.19 shows that each of the post-commissioning period calibrated simulations (Option D, MBE adjusted) has less variation (smaller range) than the corresponding regression models (Option C) for both chilled and hot water, although some of the chilled water post-commissioning periods are quite similar. The smallest chilled water range difference between Option C and Option D is 0.03 % (1997) while the largest is 1.56 % (2001). The average chilled water savings range over all post-commissioning periods for Option D (MBE adjusted) is 1.32 % while the average for Option C with regression models is 2.14 %. The hot water range differences are larger, the smallest being 1.80 % (2001) and the largest being 6.22 % (1/99-4/24/00). The mean hot water savings range over all post-commissioning periods for Option D (MBE adjusted) is 3.30 % and 7.16 % for Option C with regression models. Even though Option D (MBE adjusted) exhibits an overall lower savings range than Option C with regression models across all weather years when using the NAC weather normalization approach, the average chilled water savings from the different weather years over all post-commissioning periods are somewhat similar for Option D (MBE adjusted) and Option C with regression models. The differences for average chilled water percent savings vary from 0.48 % in 1997 to 1.81 % in 1/02-11/02. For hot water, the differences for average savings during the post-commissioning periods vary as two of the periods show differences between Option D (MBE adjusted) and Option C with regression models of less than 1 % (1997 and 1/02-11/02) and three show differences greater than 3 % (1998, 2001, and 9/03-6/04). The differences for average hot water percent savings vary from 0.54 % in 1997 up to 3.48 % in 2001.

Table 3.19 Chilled and hot water percent savings range and average across all College Station weather years under NAC weather normalization approach for Option D (MBE adjusted) and Option C with regression models.

CHW % Savings Range	1997	1998	1/99-4/24/00	2001	1/02-11/02	9/03-6/04	Average
Option D	1.30 %	0.81 %	1.49 %	1.04 %	1.42 %	1.84 %	1.32 %
Option C with Regression	1.33 %	1.82 %	1.82 %	2.59 %	2.69 %	2.60 %	2.14 %

HW % Savings Range	1997	1998	1/99-4/24/00	2001	1/02-11/02	9/03-6/04	Average
Option D	4.43 %	3.53 %	2.76 %	2.70 %	3.29 %	3.09 %	3.30 %
Option C with Regression	9.65 %	7.14 %	8.97 %	4.49 %	6.01 %	6.70 %	7.16 %

CHW % Savings Average	1997	1998	1/99-4/24/00	2001	1/02-11/02	9/03-6/04	Average
Option D	4.33 %	8.21 %	12.12 %	14.02 %	7.18 %	22.41 %	11.38 %
Option C with Regression	4.81 %	8.79 %	13.82 %	14.91 %	9.00 %	24.08 %	12.57 %

HW % Savings Average	1997	1998	1/99-4/24/00	2001	1/02-11/02	9/03-6/04	Average
Option D	18.71 %	23.50 %	25.36 %	16.04 %	-3.20 %	15.53 %	15.99 %
Option C with Regression	18.16 %	20.07 %	26.87 %	12.56 %	-4.20 %	18.87 %	15.39 %

Variability of Savings Statistical Measures

Two statistical measures are now shown to better quantify the variability of savings for each of the four different combinations of savings determination methods and weather normalization approaches (Option D using the NAC weather normalization approach, Option C with regression models using the NAC weather normalization approach, Option D using the standard IPMVP weather normalization approach, and Option C with regression models using the standard IPMVP weather normalization approach). This allows for simple comparison of savings variability between the four different combinations of sets of savings obtained from the 29 different College Station weather years and random runs for the NAC and standard IPMVP weather normalization approaches, respectively. The two statistical measures used here are the standard deviation and the coefficient of variation of the percent savings.

Chilled water savings variability results are listed in Table 3.20. Both the standard deviation and coefficient of variation results show Option D (MBE adjusted) with the NAC weather normalization approach to have the least amount of savings variability (0.32 % savings average standard deviation and 3.50 % average coefficient of variation for chilled water savings across all post-commissioning periods) while Option C with regression models using the standard IPMVP weather normalization approach has the most amount of savings variability (1.01 % savings average standard deviation and 8.36 % average coefficient of variation for chilled water savings across all post-commissioning periods). The average chilled water savings variability measures are similar for Option D (MBE adjusted) with the standard IPMVP weather normalization approach (0.68 % savings average standard deviation and 5.78 % average coefficient of variation for chilled water savings across all post-commissioning periods) and Option C with regression models with the NAC weather normalization approach (0.57 % savings average standard deviation and 5.29 % average coefficient of variation for chilled water savings across all post-commissioning periods). Thus the NAC weather normalization approach shows less variability in chilled water savings than the standard IPMVP weather normalization approach when comparing results using the same IPMVP savings determination method. Additionally, Option D (MBE adjusted) shows less variability in chilled water savings than Option C with regression models when comparing results using the same weather normalization approach.

Table 3.20 Chilled water savings variability quantification for each of the four combinations of the NAC and standard IPMVP weather normalization approaches and Option C and Option D (MBE adjusted) savings determination methods.

Option D (MBE Adjusted) with NAC Weather Normalization Approach							
	1997	1998	1/99-4/24/00	2001	1/02-11/02	9/03-6/04	Average
stdev	0.29 %	0.20 %	0.39 %	0.25 %	0.36 %	0.43 %	0.32 %
coeff var	6.76 %	2.41 %	3.19 %	1.76 %	4.95 %	1.92 %	3.50 %

Option D (MBE Adjusted) with Standard IPMVP Weather Normalization Approach							
	1997	1998	1/99-4/24/00	2001	1/02-11/02	9/03-6/04	Average
stdev	0.20 %	0.41 %	0.61 %	0.93 %	0.49 %	1.46 %	0.68 %
coeff var	4.70 %	5.06 %	5.08 %	6.57 %	6.82 %	6.44 %	5.78 %

Option C with Regression Models with NAC Weather Normalization Approach							
	1997	1998	1/99-4/24/00	2001	1/02-11/02	9/03-6/04	Average
stdev	0.35 %	0.47 %	0.48 %	0.70 %	0.73 %	0.71 %	0.57 %
coeff var	7.20 %	5.37 %	3.46 %	4.68 %	8.06 %	2.96 %	5.29 %

Option C with Regression Models with Standard IPMVP Weather Normalization Approach							
	1997	1998	1/99-4/24/00	2001	1/02-11/02	9/03-6/04	Average
stdev	0.40 %	0.69 %	0.97 %	1.27 %	1.02 %	1.74 %	1.01 %
coeff var	8.75 %	7.79 %	7.09 %	8.41 %	10.97 %	7.14 %	8.36 %

Hot water savings variability results are listed in Table 3.21. When not considering the coefficient of variation of the 1/02-11/02 where negative savings occur, hot water savings variability results are similar to those of the chilled water except that the savings variability measures are much closer for the NAC and standard IPMVP weather normalization approaches when using Option C with regression models. The average standard deviation across all post-commissioning periods is 1.83 % savings for Option C with regression models using the NAC weather normalization approach versus 1.85 % savings for Option C with regression models using the standard IPMVP weather normalization approach. When using Option D (MBE adjusted), however, the NAC weather normalization approach clearly shows less savings variability than the standard IPMVP weather normalization approach. The average standard deviation across all post-commissioning periods is 0.79 % savings for Option D (MBE adjusted) using the NAC weather normalization approach versus 1.32 % savings for Option D (MBE adjusted) using the standard IPMVP weather normalization approach. When considering the average of the coefficient of variation across all post-commissioning periods excluding the 1/02-11/02 period, the same pattern seen with the standard deviation exists. The NAC weather normalization approach (4.20 % average coefficient of variation) shows less variability in savings than the standard IPMVP weather normalization approach

(7.36 % average coefficient of variation) when using Option D (MBE adjusted) of the IPVMP. Also, the NAC weather normalization approach (9.83 % average coefficient of variation) shows similar variability in savings to the standard IPMVP weather normalization approach (9.92 % average coefficient of variation) when using Option C with regression models of the IPMVP.

Table 3.21 Hot water savings variability quantification for each of the four combinations of the NAC and standard IPMVP weather normalization approaches and Option C and Option D (MBE adjusted) savings determination methods.

Option D (MBE Adjusted) with NAC Weather Normalization Approach								
	1997	1998	1/99-4/24/00	2001	1/02-11/02	9/03-6/04	Average	Avg without 1/02-11/02
stdev	1.06 %	0.82 %	0.62 %	0.67 %	0.76 %	0.80 %	0.79 %	0.80 %
coeff var	5.69 %	3.50 %	2.46 %	4.18 %	-23.74 %	5.15 %	-0.46 %	4.20 %

Option D (MBE Adjusted) with Standard IPMVP Weather Normalization Approach								
	1997	1998	1/99-4/24/00	2001	1/02-11/02	9/03-6/04	Average	Avg without 1/02-11/02
stdev	1.14 %	1.34 %	2.11 %	1.70 %	0.69 %	0.93 %	1.32 %	1.44 %
coeff var	6.05 %	5.62 %	8.39 %	10.76 %	-21.82 %	5.96 %	2.49 %	7.36 %

Option C with Regression Models with NAC Weather Normalization Approach								
	1997	1998	1/99-4/24/00	2001	1/02-11/02	9/03-6/04	Average	Avg without 1/02-11/02
stdev	2.54 %	1.80 %	2.09 %	1.13 %	1.61 %	1.78 %	1.83 %	1.87 %
coeff var	14.00 %	8.95 %	7.79 %	8.96 %	-38.48 %	9.46 %	1.78 %	9.83 %

Option C with Regression Models with Standard IPMVP Weather Normalization Approach								
	1997	1998	1/99-4/24/00	2001	1/02-11/02	9/03-6/04	Average	Avg without 1/02-11/02
stdev	2.49 %	1.72 %	2.38 %	1.58 %	1.70 %	1.22 %	1.85 %	1.88 %
coeff var	13.45 %	8.31 %	8.91 %	12.60 %	-47.54 %	6.32 %	0.34 %	9.92 %

Variability of Persistence of Savings Statistical Measures

Just as it is important to have some quantifiable measure of the variability of commissioning savings for each of the four different combinations of savings determination methods and weather normalization approaches, it is also valuable to quantify the variability of commissioning persistence of savings for each of these four different combinations. Persistence of savings in this case is defined as the percent savings difference between a post-commissioning period after 1997 (the first post-commissioning period) and the 1997 post-commissioning period.

Table 3.22 shows the chilled water savings and persistence of savings for both the NAC and standard IPMVP weather normalization approaches using Option D (MBE adjusted) of the IPMVP. The long-term average College Station weather year is used for the NAC weather normalization approach. The normal procedure for the standard IPMVP weather normalization approach is employed that uses the actual weather data of the post-commissioning periods. Table 3.22 indicates that persistence of savings results do vary depending on the weather normalization approach used just as savings results do. For two of the post-commissioning periods (1/99-4/24/00 and 1/02-11/02), chilled water persistence of savings differs by more than 1.5 % between the NAC and standard IPMVP weather normalization approaches.

Table 3.22 Chilled water savings and persistence of savings results using Option D (MBE adjusted) of IPMVP for both NAC and standard IPMVP weather normalization approaches. The long-term average weather year is used for the NAC weather normalization approach.

	1997	1998	1/99-4/24/00	2001	1/02-11/02	9/03-6/04
NAC Weather Normalization Approach						
CHW % Savings	4.62 %	8.28 %	12.24 %	13.87 %	6.91 %	22.21 %
CHW Persistence	n/a	3.66 %	7.62 %	9.25 %	2.29 %	17.59 %

	1997	1998	1/99-4/24/00	2001	1/02-11/02	9/03-6/04
Standard IPMVP Weather Normalization Approach						
CHW % Savings	4.05 %	8.18 %	9.99 %	13.43 %	8.06 %	22.06 %
CHW Persistence	n/a	4.13 %	5.94 %	9.38 %	4.01 %	18.01 %

	1997	1998	1/99-4/24/00	2001	1/02-11/02	9/03-6/04
CHW % Savings Difference	0.57 %	0.10 %	2.25 %	0.44 %	1.15 %	0.15 %
CHW Persistence Difference	n/a	0.47 %	1.68 %	0.13 %	1.72 %	0.42 %

To quantify the variability of persistence of savings, sets of savings from the 29 different College Station weather years and 29 different random runs are again used for the NAC and standard IPMVP weather normalization approaches, respectively. From these sets of persistence of savings results, the standard deviation and the coefficient of variation of the persistence are found. Table 3.23 quantifies the chilled water variability of persistence of savings. Results of the variability of chilled water persistence of savings differ somewhat from the results of the variability of chilled water savings. Option C with regression models using the NAC weather normalization approach (0.38 % persistence average standard deviation and 5.12 % average coefficient of variation across all post-commissioning periods) shows slightly less overall variability than Option D (MBE adjusted) using the NAC weather normalization approach (0.42 % persistence average standard deviation and 7.37 % average coefficient of variation across all post-commissioning periods). Option D (MBE adjusted) with the standard IPMVP weather normalization approach (0.68 % persistence average standard deviation and 9.04 % average coefficient of variation across all post-commissioning periods), however, still shows less variability than Option C with regression models using the standard IPMVP weather normalization approach (1.20 % persistence average standard deviation and 15.03 % average coefficient of variation across all post-commissioning periods). As with chilled water variability of savings results, the NAC weather normalization approach shows less chilled water variability of persistence than the standard IPMVP weather normalization approach when comparing results using the same IPMVP savings determination method (both Option C with regression models and Option D).

Table 3.23 Chilled water persistence of savings variability quantification for each of the four combinations of the NAC and standard IPMVP weather normalization approaches and Option C and Option D (MBE adjusted) savings determination methods.

Option D (MBE Adjusted) with NAC Weather Normalization Approach						
	1998	1/99-4/24/00	2001	1/02-11/02	9/03-6/04	Average
stdev	0.17 %	0.22 %	0.42 %	0.61 %	0.70 %	0.42 %
coeff var	4.38 %	2.84 %	4.29 %	21.49 %	3.87 %	7.37 %

Option D (MBE Adjusted) with Standard IPMVP Weather Normalization Approach						
	1998	1/99-4/24/00	2001	1/02-11/02	9/03-6/04	Average
stdev	0.30 %	0.46 %	0.80 %	0.45 %	1.38 %	0.68 %
coeff var	7.78 %	6.02 %	8.21 %	15.65 %	7.53 %	9.04 %

Option C with Regression Models with NAC Weather Normalization Approach						
	1998	1/99-4/24/00	2001	1/02-11/02	9/03-6/04	Average
stdev	0.18 %	0.17 %	0.44 %	0.49 %	0.60 %	0.38 %
coeff var	4.61 %	1.85 %	4.37 %	11.68 %	3.10 %	5.12 %

Option C with Regression Models with Standard IPMVP Weather Normalization Approach						
	1998	1/99-4/24/00	2001	1/02-11/02	9/03-6/04	Average
stdev	0.84 %	1.01 %	1.38 %	1.07 %	1.71 %	1.20 %
coeff var	19.39 %	11.12 %	13.07 %	22.91 %	8.65 %	15.03 %

When assessing hot water variability of persistence of savings results (see Table 3.24), the coefficient of variation should not be considered. Due to very small average persistence values in multiple post-commissioning periods, many of the results for the coefficient of variation cannot accurately assess the variability of hot water persistence of savings. The standard deviation appears better suited as a measure of variability of hot water persistence of savings. When only considering the average value across all post-commissioning periods of the standard deviation for hot water persistence of savings, Option D (MBE adjusted) using the NAC weather normalization approach has the least variability (0.98 % persistence average standard deviation across all post-commissioning periods). This is followed by Option D (MBE adjusted) using the standard IPMVP weather normalization approach (1.12 % persistence average standard deviation across all post-commissioning periods), Option C with regression models using the NAC weather normalization approach (1.75 % persistence average standard deviation across all post-commissioning periods), and Option C with regression models using the standard IPMVP weather normalization approach (2.61 % persistence average standard deviation across all post-commissioning periods).

Table 3.24 Hot water persistence of savings variability quantification for each of the four combinations of the NAC and standard IPMVP weather normalization approaches and Option C and Option D (MBE adjusted) savings determination methods.

Option D (MBE Adjusted) with NAC Weather Normalization Approach						
	1998	1/99-4/24/00	2001	1/02-11/02	9/03-6/04	Average
stdev	0.32 %	1.09 %	1.24 %	1.69 %	0.56 %	0.98 %
coeff var	6.78 %	16.43 %	-46.25 %	-7.71 %	-17.61 %	-9.67 %

Option D (MBE Adjusted) with Standard IPMVP Weather Normalization Approach						
	1998	1/99-4/24/00	2001	1/02-11/02	9/03-6/04	Average
stdev	0.80 %	1.59 %	1.16 %	1.61 %	0.44 %	1.12 %
coeff var	16.30 %	25.21 %	-37.33 %	-7.30 %	-12.98 %	-3.22 %

Option C with Regression Models with NAC Weather Normalization Approach						
	1998	1/99-4/24/00	2001	1/02-11/02	9/03-6/04	Average
stdev	0.83 %	1.31 %	2.10 %	2.98 %	1.51 %	1.75 %
coeff var	43.68 %	15.09 %	-37.40 %	-13.31 %	214.89 %	44.59 %

Option C with Regression Models with Standard IPMVP Weather Normalization Approach						
	1998	1/99-4/24/00	2001	1/02-11/02	9/03-6/04	Average
stdev	2.42 %	3.03 %	2.36 %	3.25 %	1.98 %	2.61 %
coeff var	111.63 %	37.05 %	-39.69 %	-14.71 %	271.96 %	73.25 %

Summary Comparison of Savings Using Four Different Normalization Approaches

The variability of savings and persistence of savings results from the commissioning of CE/TTI using the NAC and standard IPMVP weather normalization approaches, as well as Option C with regression models and Option D of the IPMVP are presented in this chapter. It has been shown that the savings and persistence of savings may vary greatly depending on which weather normalization approach, IPMVP Option, and weather year selected as the "normal" weather year are used.

Overall, the NAC weather normalization approach shows less variability in savings and persistence of savings than the standard IPMVP weather normalization approach. Additionally, Option D of the IPMVP generally shows less variability in savings and persistence of savings than Option C with regression models. These statements hold true when considering chilled water savings. However, for hot water savings, results for the variability in the persistence of savings are mixed. Chilled water persistence of savings results differ in that Option C with regression models using the NAC weather normalization approach shows slightly less overall variability of persistence than Option D using the NAC weather normalization approach. Finally, the hot water persistence of savings shows mixed variability.

For chilled water savings, the average standard deviation across all post-commissioning periods is 0.32 % savings for Option D (MBE adjusted) using the NAC weather normalization approach, 0.68 % savings for Option D (MBE adjusted) using the standard IPMVP weather normalization approach, 0.57 % savings for Option C with regression models using the NAC weather normalization approach, and 1.01 % savings for Option C with regression models using the standard IPMVP weather normalization approach.

For hot water savings, the average standard deviation across all post-commissioning periods is 0.79 % savings for Option D (MBE adjusted) using the NAC weather normalization approach, 1.32 % savings for Option D (MBE adjusted) using the standard IPMVP weather normalization approach, 1.83 % savings for Option C with regression models using the NAC weather normalization approach, and 1.85 % savings for Option C with regression models using the standard IPMVP weather normalization approach.

For chilled water persistence of savings, the average standard deviation across all post-commissioning periods is 0.42 % persistence for Option D (MBE adjusted) using the NAC weather normalization approach, 0.68 % persistence for Option D (MBE adjusted) using the standard IPMVP weather normalization approach, 0.38 % persistence for Option C with regression models using the NAC weather normalization approach, and 1.20 % persistence for Option C with regression models using the standard IPMVP weather normalization approach.

For hot water persistence of savings, the average standard deviation across all post-commissioning periods is 0.98 % persistence for Option D (MBE adjusted) using the NAC weather normalization approach, 1.12 % persistence for Option D (MBE adjusted) using the standard IPMVP weather normalization approach, 1.75 % persistence for Option C with regression models using the NAC weather normalization approach, and 2.61 % persistence for Option C with regression models using the standard IPMVP weather normalization approach.

3.4 Recent Persistence Studies

3.4.1 Persistence of Commissioning Measures in Four Buildings in Japan

Yamaha (2007) examined information available in Japanese for 14 buildings that had received "10 Year Awards" from the Society of Heating, Air-Conditioning and Sanitary Engineers of Japan (SHASE) for conducting system inspections and improvements over a period of at least ten years. He concluded there was sufficient information regarding energy-saving measures implemented and subsequent energy use in three of these buildings to be valuable for examining the persistence of energy savings in buildings. In these cases, the measures were primarily energy efficiency measures incorporated into the design of the buildings.

He discussed the Chubu Electric Power Co. (CEPCO) Okazaki Building, the Tokyo Electric Power Co. (TEPCO) Higashi-Murayama Building and the TEPCO R&D Center, all of which received SHASE Ten Year Awards. These buildings used 20 % to 40 % less energy than reference designs. He also presents the performance of the Tonets Corp. (TONETS) Shinkawa Building.

The CEPCO Okazaki Building incorporated high levels of insulation, sun shading with controls and cubic shape to minimize surface area. The building also uses thermal energy storage, variable water volume and variable air volume control strategies and heat recovery. Figure 3.10 shows the disaggregated energy consumption of the CEPCO Okazaki Building for 20 years. The energy use ranges from a low of 50 % of the reference consumption to a high of about 80 % of the reference consumption. During the latter half of the 1980s, building consumption increased substantially, primarily due to increases in lighting consumption, fan energy and pumping power. But even after these increases, the building still used 20 % less than the reference design.

The TEPCO Higashi-Murayama Building incorporates ventilation windows, sun shade controls, data visualization tools, thermal energy storage and large temperature differences in the distribution systems. As shown in Fig. 3.11, the consumption was relatively stable over the last 11 years shown, with consumption between a low of about 75 % of the reference building (in a year with a cool summer) to a high of about 85 % of the reference. Consumption was very low in 1988 since the building was not operational for the entire year.

The TEPCO R&D Center utilizes data visualization tools to help achieve optimal operation and incorporates thermal storage and large temperature differences in the distribution systems. The energy consumption shown in Fig. 3.12 is quite stable for four of the nine years shown, has three years of abnormally high consumption (1997 to 1999) due to increased consumption of an experimental fuel cell installation in the building. During 2002 and 2003 consumption decreased by 15 % and 26 % due to the removal of the experimental fuel cell installation and a cool summer in 2003.

Figure 3.13 shows the savings in MJ/m^2-year of the CEPCO Okazaki Building, the TEPCO Higashi-Murayama Building and the TONETS Shinkawa Building. The first two saved significant energy throughout the entire period of observation along with the TEPCO R&D building. The TONETS Shinkawa Building saved significant energy for seven years, then over the next seven years shows savings that are negative in aggregate due to a 25 % increase in operating hours and additional computers and office machines.

As a group, these buildings are consistent with the commissioned buildings that have been examined in Annex 47 work – most continue to show savings over periods that in this case run from 10 years to 20 years, while generally showing some decrease in savings over time.

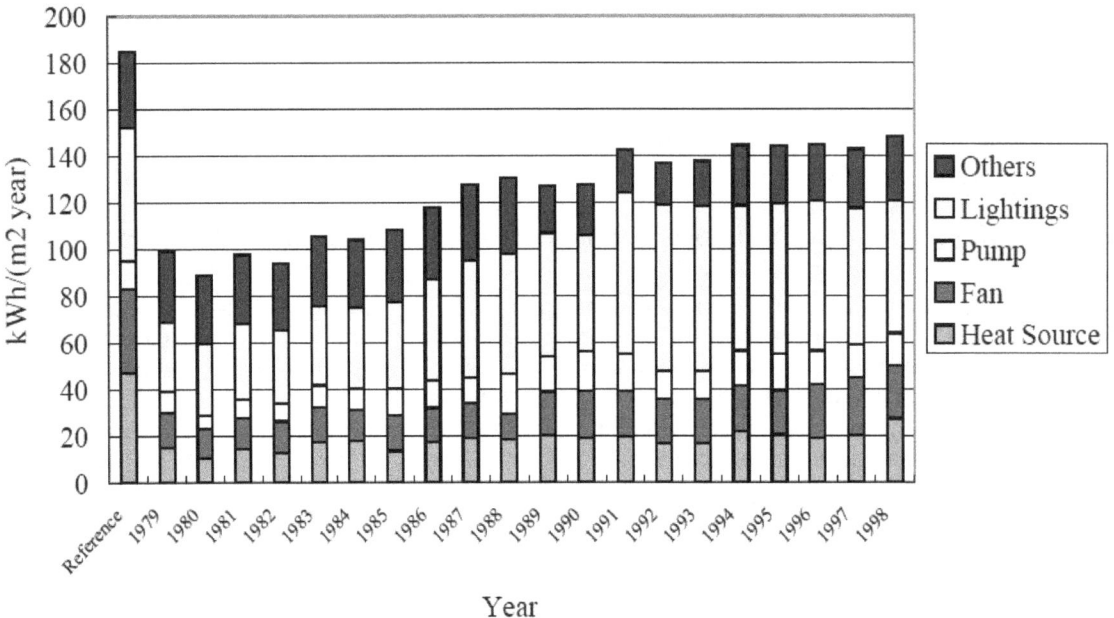

Figure 3.10 Annual energy consumption of the CEPCO Okazaki Building for years 1979 to 1998 (Source: Yamaha 2007).

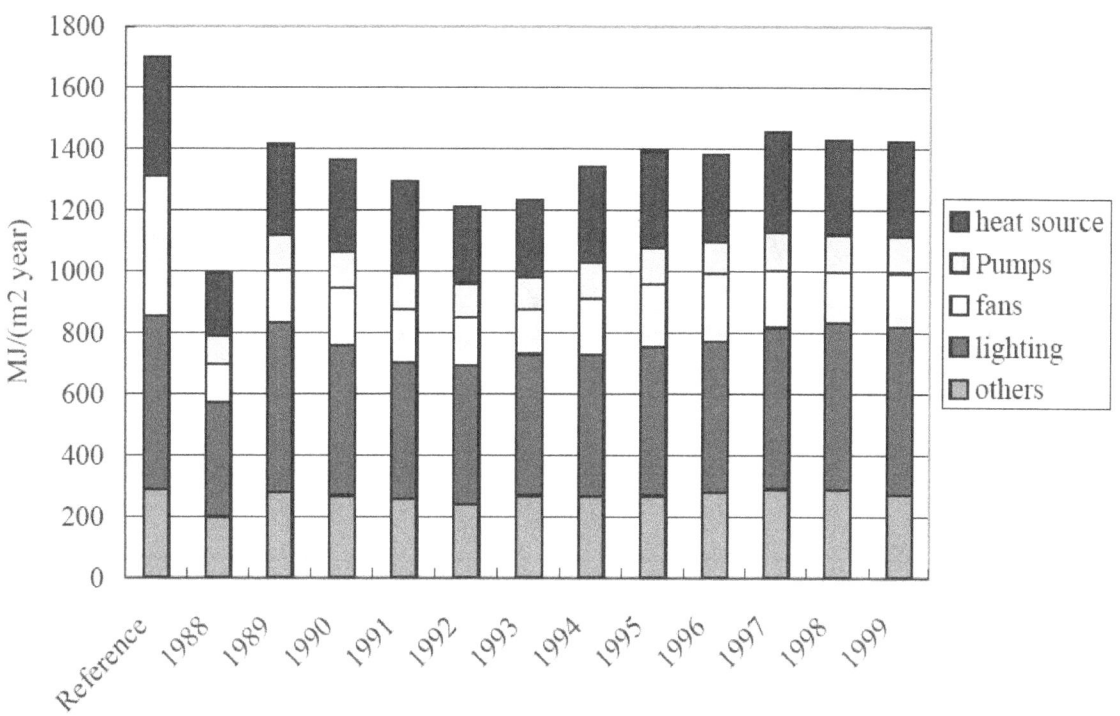

Figure 3.11 Energy consumption of the TEPCO Higashi Murayama Building for 1988 to 1999 (Source: Yamaha 2007).

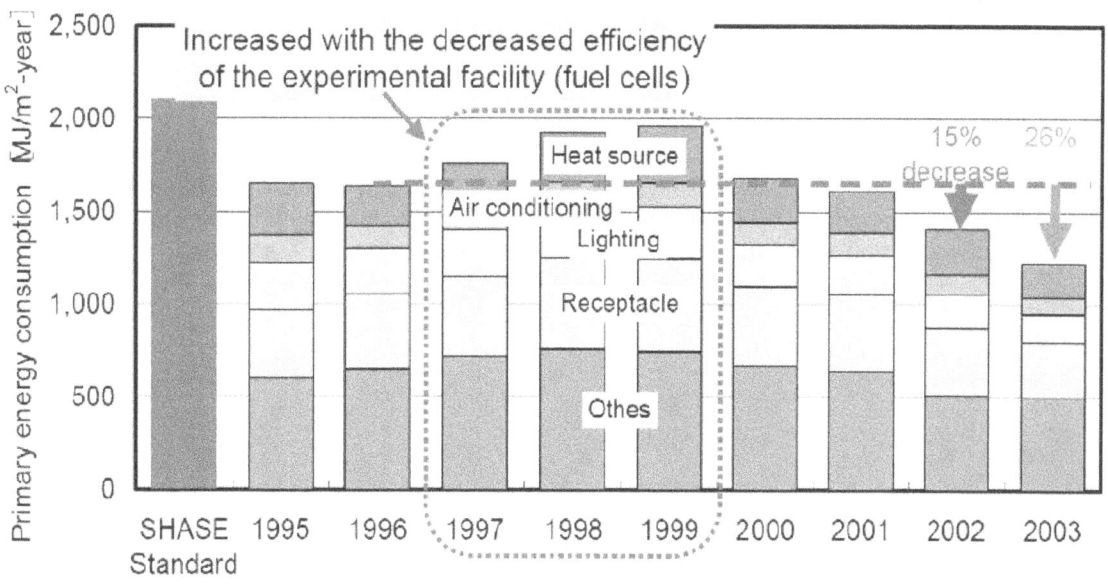

Figure 3.12 Energy consumption of the TEPCO R&D for 1995 to 2003 (Source: Yamaha 2007).

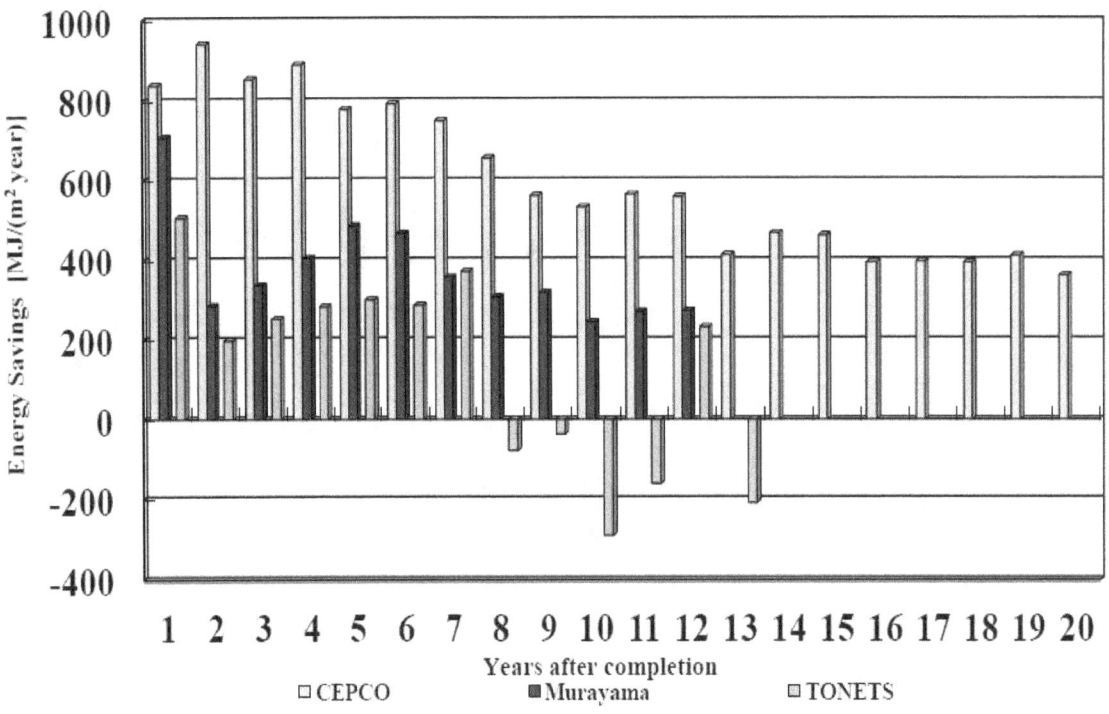

Figure 3.13. Persistence of energy savings for three buildings over a 20 year period (Source: Yamaha 2007).

3.4.2 Persistence of Commissioning Measures in ten Buildings at Texas A&M – An Update

Results of Previous Study

As noted earlier, the previous study of the ten buildings by Turner et al. (2001) and Cho (2002) compared the normalized energy savings of each building over a period of four years following retro-commissioning. These results have been updated by Toole (2009). Table 3.25 details the results of this study, with the chilled water, hot water, and electricity consumption and savings shown on a yearly basis.

Table 3.25 Energy savings results for the years examined in the previous study (Cho, 2002).

Building Name	Type	Baseline Use (MMBtu) (MWh) / yr	1997 Use (MMBtu) (MWh) / yr	1997 Saving (%)	1998 Use (MMBtu) (MWh) / yr	1998 Saving (%)	1999 Use (MMBtu) (MWh) / yr	1999 Saving (%)	2000 Use (MMBtu) (MWh) / yr	2000 Saving (%)
Blocker	CHW	22,955	16,723	27	19,530	15	20,164	12	21,083	** 8
	HW	8,735	4,093	53	1,676	81	3,330	62	4,344	** 50
	Elec	4,832	3,773	22	3,883	20	3,936	19	3,859	20
Eller O&M	CHW	30,625	18,846	38	18,660	39	19,012	38	20,360	34
	HW	7,584	2,578	66	1,154	85	1,831	76	4,712	38
	Elec	4,891	3,698	24	3,675	25	3,823	22	3,874	21
G.R.White Coliseum	CHW	18,872	8,717	54	8,511	55	14,548	23	15,858	16
	HW	21,155	6,091	71	549	97	4,923	77	10,111	52
	Elec	1,480	1,297	12	1,168	21	1,171	21	1,291	13
Harrington Tower	CHW	14,179	7,109	50	8,420	41	7,660	46	9,032	36
	HW	6,896	2,603	62	914	87	1,629	76	3,519	49
	Elec	1,666	1,297	22	1,336	20	1,341	20	1,353	19
Kleberg Building	CHW	59,271	34,864	41	34,969	41	36,731	38	41,965	29
	HW	40,812	6,523	84	1,215	97	8,030	80	10,591	74
	Elec	5,511	5,458	1	5,067	8	4,778	13	4,684	15
Koldus Building	CHW	* 21,964	12,177	45	12,988	41	12,740	42	11,804	46
	HW	2,103	704	67	399	81	634	70	649	69
	Elec	2,850	2,511	12	2,597	9	2,624	8	2,592	9
Rich. Petroleum	CHW	28,526	13,599	52	15,637	45	15,078	47	17,702	38
	HW	* 18,227	6,565	64	5,588	69	5,098	72	2,171	88
	Elec	1,933	1,898	2	1,914	1	1,991	-3	2,153	-11
VMC Addition	CHW	40,892	23,115	43	24,080	41	22,915	44	23,307	43
	HW	3,569	887	75	2,041	43	2,097	41	2,051	43
	Elec	4,186	3,996	5	4,140	1	4,236	-1	4,056	3
Wehner CBA	CHW	19,193	12,327	36	13,339	31	12,530	35	11,609	40
	HW	13,393	10,876	19	9,715	27	6,581	51	6,350	53
	Elec	2,555	2,410	6	2,446	4	2,552	0	2,581	-1
Zachry Engr. Center	CHW	40,824	16,737	59	17,377	57	18,148	56	20,225	50
	HW	7,676	1,630	79	3,230	58	2,226	71	4,271	44
	Elec	7,502	6,762	10	6,793	9	7,099	5	6,955	7
Type		Total	Total	Average	Total	Average	Total	Average	Total	Average
Chilled Water		297,298	164,215	44.8	173 509	41.6	179,527	39.6	192,946	35.1
Hot Water		130,149	42,549	67.3	26,482	79.7	36,380	72.0	48,768	62.5
Electricity		37,407	33,100	11.5	33 018	11.7	33,552	10.3	33,399	10.7

* The baseline energy use for these buildings was estimated from the average savings of the other buildings because insufficient data was available to create reliable baselines.

** The Blocker building had insufficient chilled water and hot water energy use data in 2000 to determine normalized annual consumption. So the savings were estimated from the average degradation that occurred between 1999 and 2000 in the other nine buildings.

New Findings

The results of the previous study were expanded upon to include normalized consumption data and savings calculations for additional years following the completion of the original study. For seven of the buildings, reliable energy consumption data were available from as recently as 2006-2007. For the other three buildings, the last year of reliable consumption data ranged from 2002 to 2004. Table 3.26 shows the combined results of the previous study with the additional years of data for each building.

Table 3.26 Updated results of energy savings analysis, normalized to common weather year.

Building Name	Type	Baseline Use (MMBtu) (MWh)/yr	1997 Use (MMBtu) (MWh)/yr	1997 Saving (%)	1998 Use (MMBtu) (MWh)/yr	1998 Saving (%)	1999 Use (MMBtu) (MWh)/yr	1999 Saving (%)	2000 Use (MMBtu) (MWh)/yr	2000 Saving (%)	2001 Use (MMBtu) (MWh)/yr	2001 Saving (%)	2002 Use (MMBtu) (MWh)/yr	2002 Saving (%)	2003 Use (MMBtu) (MWh)/yr	2003 Saving (%)	2004 Use (MMBtu) (MWh)/yr	2004 Saving (%)	2005-2006 Use (MMBtu) (MWh)/yr	2005-2006 Saving (%)	2006-2007 Use (MMBtu) (MWh)/yr	2006-2007 Saving (%)
Blocker	CHW	22,955	16,723	27	19,530	15	20,164	12	21,063	8 **	19,082	17	17,887	22	20,850	9					21,179	8
	HW	8,735	4,093	53	1,676	81	3,330	62	4,344	50 **	4,623	47	2,654	70	6,367	27					4,409	50
	Elec	4,832	3,773	22	3,883	20	3,936	19	3,859	20	3,639	25	3,596	27	3,583	26					3,511	27
Eller O&M	CHW	30,625	18,846	38	18,660	39	19,012	38	20,360	34	24,002	22	21,120	31	19,948	35	21,805	29				
	HW	7,584	2,578	66	1,154	85	1,831	76	4,712	38	4,488	41										
	Elec	4,891	3,698	24	3,675	25	3,823	22	3,874	21	3,972	19	3,732	24	3,745	23	3,861	21				
G.R. White Coliseum	CHW	18,872	8,717	54	8,511	55	14,548	23	15,858	16									6,837	64	11,134	41
	HW	21,155	6,091	71	549	97	4,923	77	10,111	52									3,276	85	2,216	90
	Elec	1,480	1,297	12	1,168	21	1,171	21	1,291	13	1,102	26	1,028	31	1,015	31	1,109	25	1,028	31	956	35
Harrington Tower	CHW	14,179	7,109	50	8,420	41	7,660	46	9,032	36	8,380	41	9,267	35	8,614	39	7,817	45	7,103	50	6,927	51
	HW	6,896	2,603	62	914	87	1,629	76	3,519	49			3,921	43	3,538	49			2,966	57	2,807	59
	Elec	1,666	1,297	22	1,336	20	1,341	20	1,353	19	1,319	21	1,331	20	1,390	17			1,293	22	1,220	27
Kleberg Building	CHW	59,271	34,864	41	34,969	41	36,731	38	41,965	29	45,187	24	37,180	37	31,911	46	33,560	43	20,964	65	28,831	51
	HW	40,812	6,523	84	1,215	97	8,030	80	10,561	74									7,421	82	12,969	68
	Elec	5,511	5,458	1	5,067	8	4,778	13	4,684	15	4,539	18	4,564	17	4,832	12	4,666	15	3,320	40	3,533	36
Koldus Building	CHW	21,964	12,177	45	12,998	41	12,740	42	11,804	46	12,735	42							12,487	43	13,784	37
	HW	2,103	704	67	399	81	634	70	649	69	390	81							3,488	-66	4,225	-101
	Elec	2,850	2,511	12	2,597	9	2,624	8	2,592	9	2,603	9	2,667	6			2,682	6	2,553	10	2,546	11
Rich. Petroleum	CHW	28,526	13,599	52	15,637	45	15,078	47	17,702	38	13,937	51	15,587	45	17,023	40	17,625	38				
	HW	18,227	6,565	64	5,588	69	5,098	72	2,171	88	6,568	64	6,994	62	7,391	59	8,882	51				
	Elec	1,933	1,898	2	1,914	1	1,991	-3	2,153	-11	2,039	-5	2,026	-5	2,110	-9	2,155	-11				
VMC Addition	CHW	40,892	23,115	43	24,060	41	22,915	44	23,307	43	24,360	40	25,849	37								
	HW	3,569	887	75	2,041	43	2,097	41	2,051	43	1,881	47	3,208	10								
	Elec	4,186	3,996	5	4,140	1	4,236	-1	4,066	3	4,219	-1	4,169	0								
Wehner CBA	CHW	19,193	12,327	36	13,339	31	12,530	35	11,609	40	13,490	30									13,283	31
	HW	13,393	10,876	19	9,715	27	6,581	51	6,350	53	7,309	45									1,723	87
	Elec	2,555	2,410	6	2,446	4	2,552	0	2,561	-1	2,529	1									2,342	8
Zachry Engr. Center	CHW	40,824	16,737	59	17,377	57	18,148	56	20,225	50	19,794	52							20,440	50	24,296	40
	HW	7,676	1,630	79	3,230	58	2,226	71	4,271	44	4,467	44							3,623	53	4,694	39
	Elec	7,502	6,762	10	6,793	9	7,099	5	6,955	7	6,597	12	6,516	13	6,456	14			4,377	42	4,662	38

Note: The consumption data used for the time period labeled "2005-2006" were from 7/25/2005 – 7/24/2006 for all of the buildings with data for this period. For the period labeled "2006-2007," the consumption data were from 7/25/2006 – 7/24/2007 for G.R. White Coliseum, Harrington Tower, and Kleberg, and were from 10/16/2006 – 10/15/2007 for Blocker, Wehner, and Zachry. These time periods were chosen due to the availability of reliable energy consumption data.

The overall trends in chilled water, hot water, and electricity savings over the period sampled for the ten buildings are diagrammed in Figures 3.14, 3.15, and 3.16 that follow. Specifics about the savings patterns of each building are discussed thereafter.

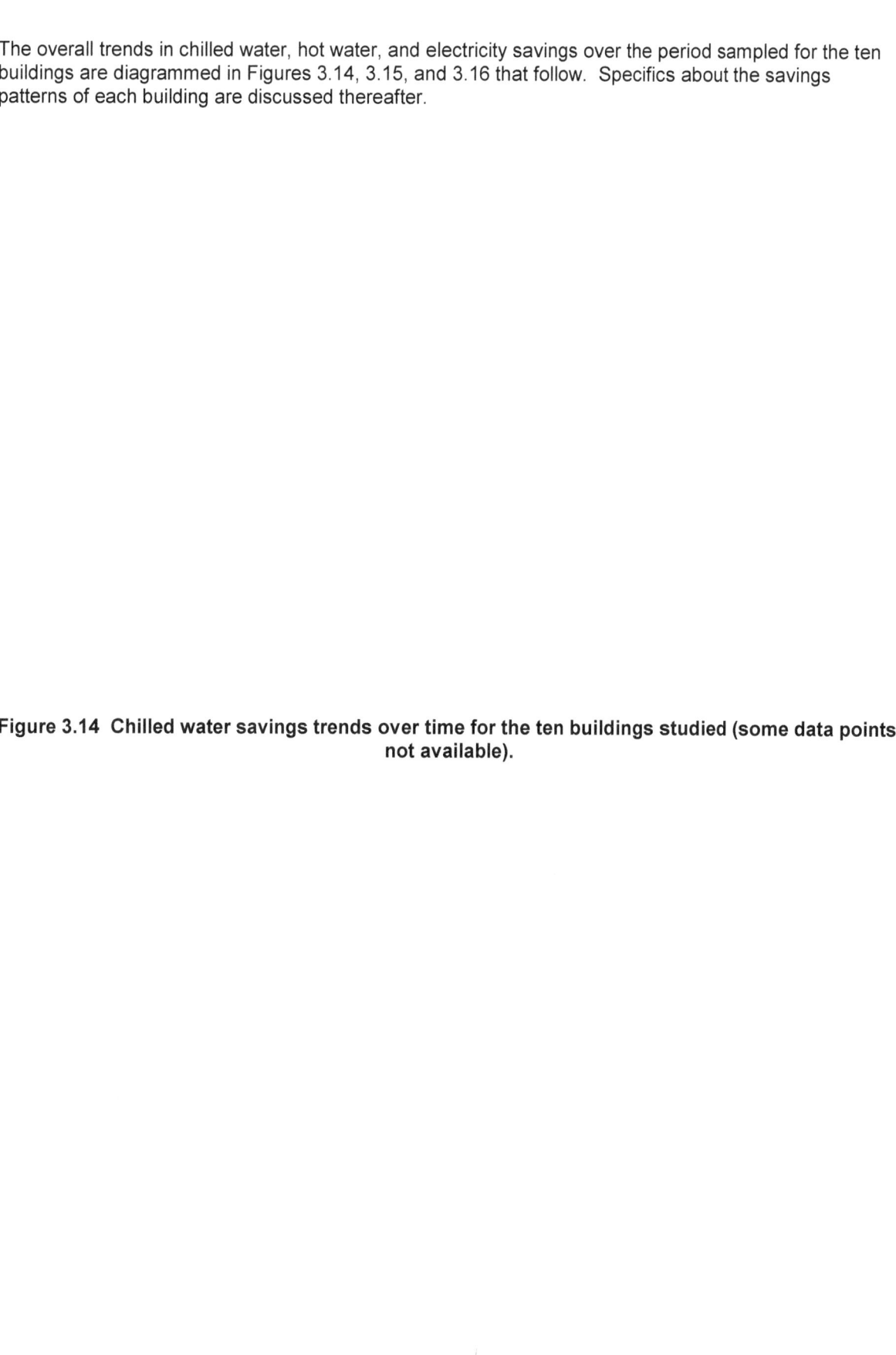

Figure 3.14 Chilled water savings trends over time for the ten buildings studied (some data points not available).

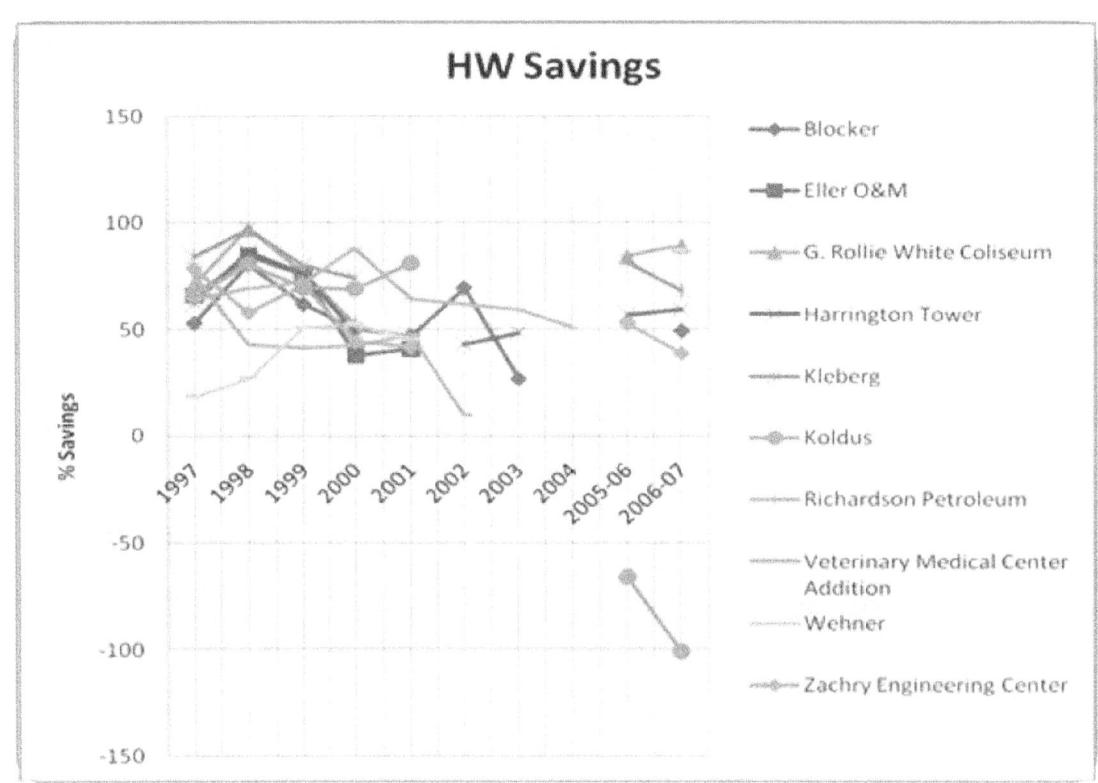

Figure 3.15 Hot water savings trends over time for the ten buildings studied.

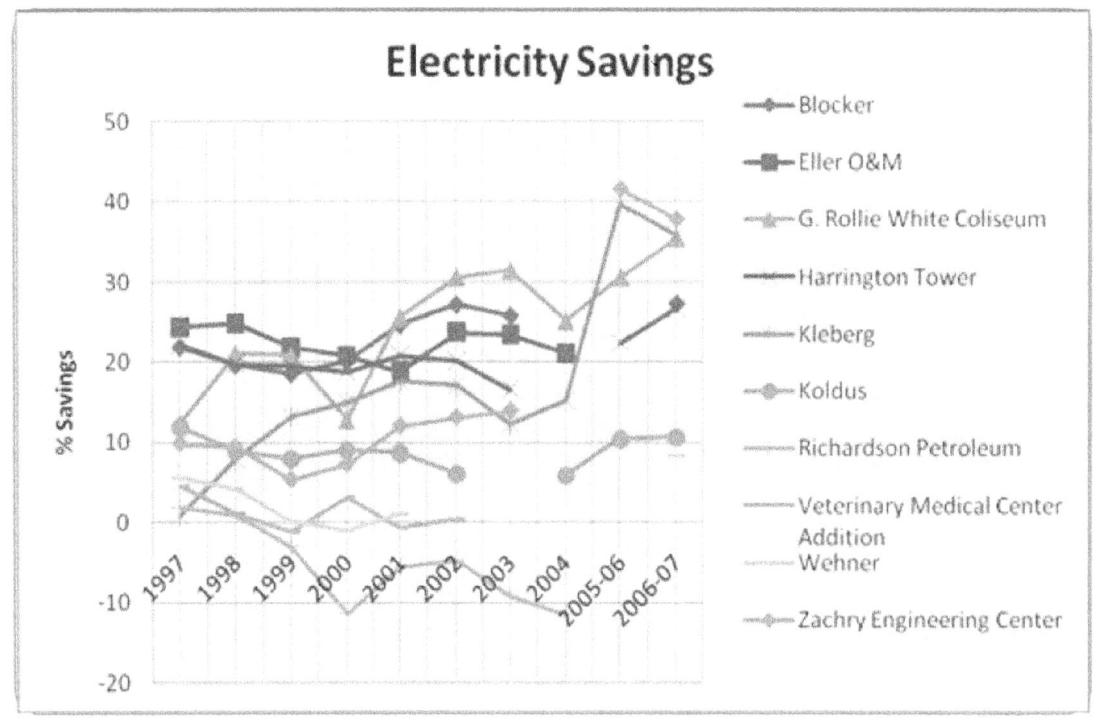

Figure 3.16 Electricity savings trends over time for the ten buildings studied.

Blocker

The savings trends for chilled water, hot water, and electricity consumption for the Blocker Building are shown in bar graph form in Fig. 3.17.

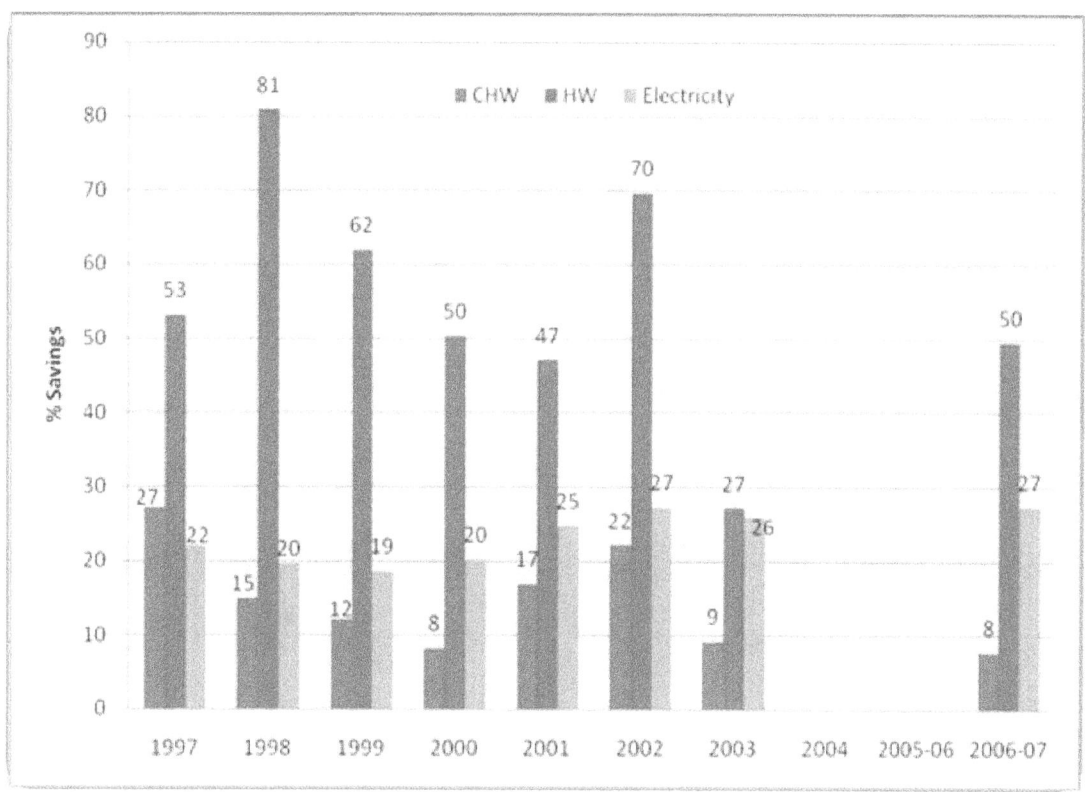

Figure 3.17 Normalized energy savings patterns for the Blocker Building.

The chilled water savings achieved in 1997 was 27 %, but had degraded to 8 % by 2000. However, over the next two years the chilled water savings increased, then fell again, so that by 2006-07 it had again fallen to 8 %. The hot water savings achieved in 1997 was 53 %, and in 2006-07 was a close 50 %. During the years between, however, it rose as high as 81 %, while dropping as low as 27 %. The electricity savings remained fairly constant in the ten year period, even rising some from 22 % in 1997 to 27 % in 2006-07.

Eller O&M

The savings trends for chilled water, hot water, and electricity consumption for the Eller O&M Building are shown in bar graph form in Fig. 3.18.

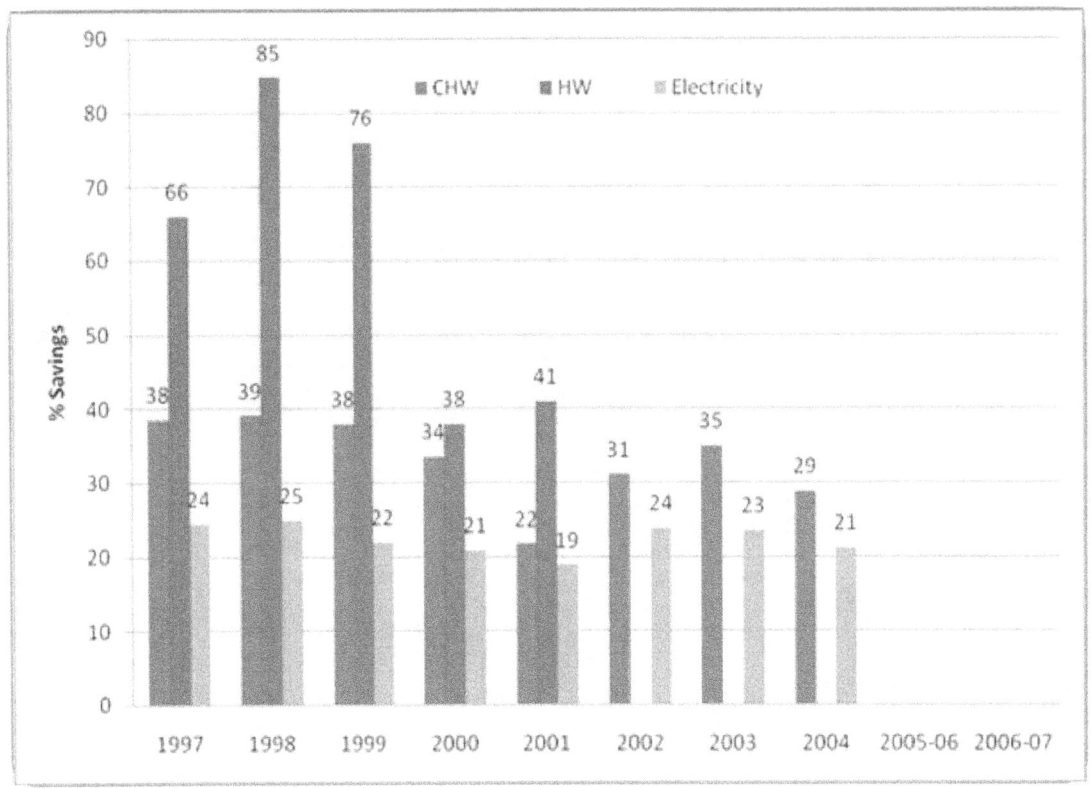

Figure 3.18 Normalized energy savings patterns for the Eller O&M Building.

The chilled water savings achieved in 1997 was 38 %. By 2004, the last year of available data, the savings had degraded slightly to 29 %. The hot water savings achieved in 1997 was 66 %, increased to 85 % the next year and 76 % the next, and then declined sharply to 38 % and 41 % in the final two years of available data. The electricity savings remained fairly constant in the eight year period of available data, beginning at 24 % in 1997 and falling slightly to 21 % by 2004.

G. Rollie White Coliseum

The savings trends for chilled water, hot water, and electricity consumption for the G. Rollie White Coliseum are shown in bar graph form in Fig. 3.19.

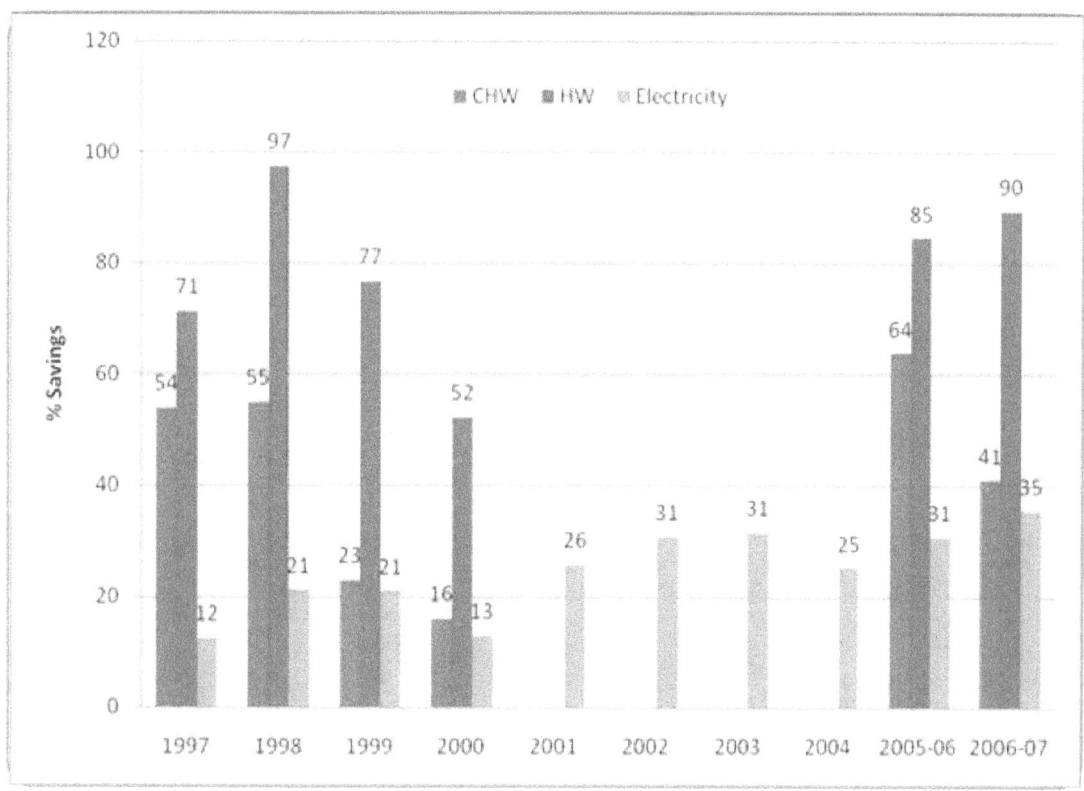

Figure 3.19 Normalized energy savings patterns for the G. Rollie White Coliseum.

The G. Rollie White Coliseum experienced some rather dramatic swings in the level of savings in both chilled water and hot water consumption, particularly in the first few years after retro-commissioning. However, by the later years (2005-2007), the level of savings for hot water was close to or exceeded previous savings, and the level of chilled water savings had risen again and settled out at a level of 41 %, or 13 % lower than the 1997 value. The electricity savings actually increased fairly steadily over time, rising from 12 % in 1997 to 35 % in 2006-07.

Harrington Tower

The savings trends for chilled water, hot water, and electricity consumption for the Harrington Tower are shown in bar graph form in Fig. 3.20.

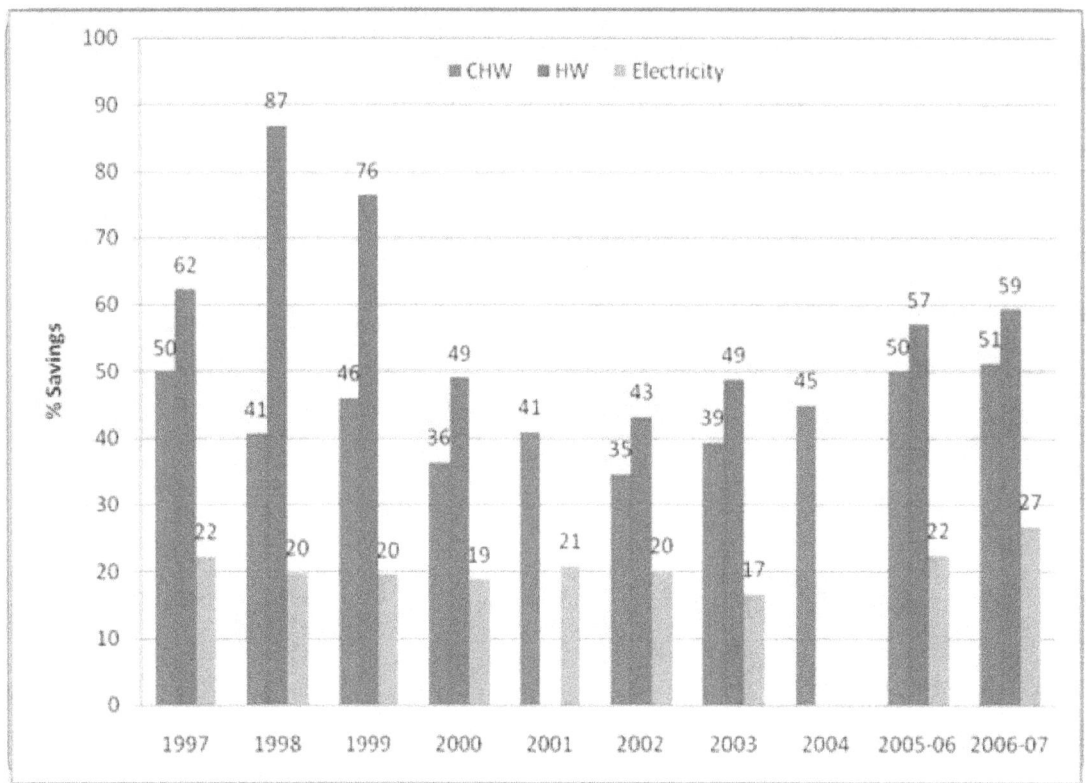

Figure 3.20 Normalized energy savings patterns for the Harrington Tower.

Harrington Tower demonstrated remarkable levels of savings persistence in both chilled water and electricity consumption, actually increasing slightly in the level of savings of each in a ten year period. While the hot water savings ended up considerably lower than the peak level achieved (down to 59 % from 87 %), it had risen in later years, and ended very close to the level achieved originally in 1997 (62 %).

Kleberg

The savings trends for chilled water, hot water, and electricity consumption for the Kleberg Building are shown in bar graph form in Fig. 3.21.

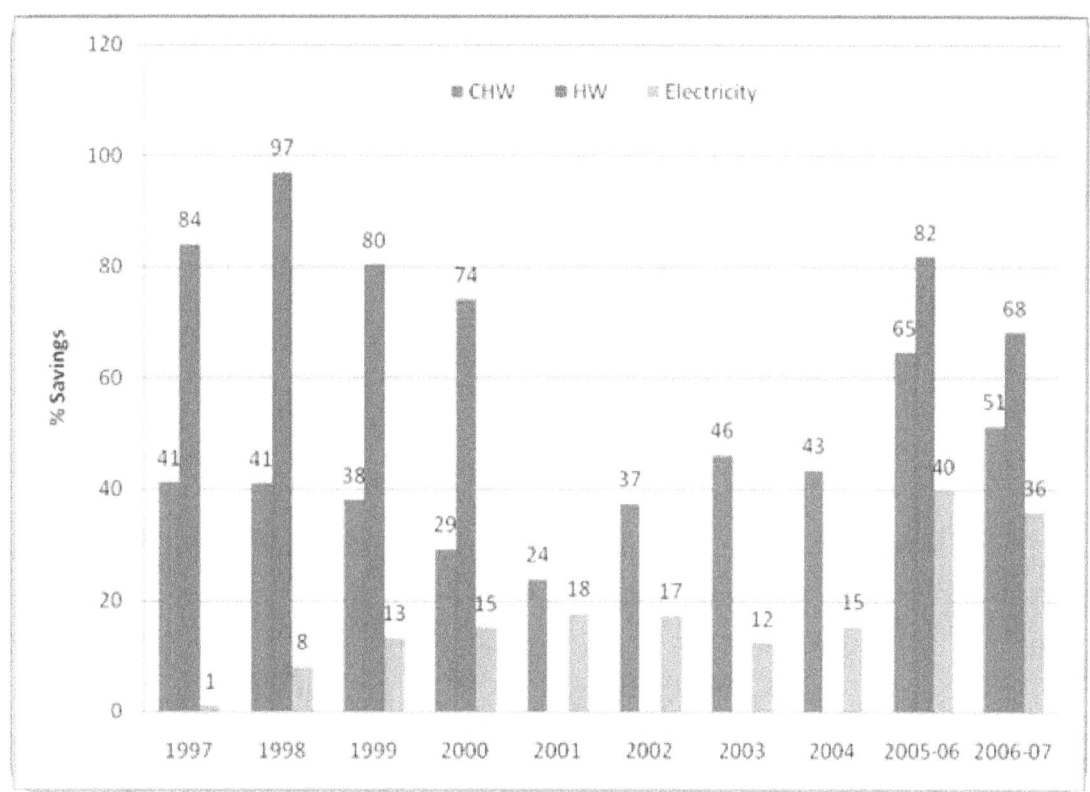

Figure 3.21 Normalized energy savings patterns for the Kleberg Building.

The level of hot water savings in Kleberg decreased from a peak of 97 % in 1998 to 68 % in 2006-07. However, the chilled water savings increased during the same period, rising from 41 % in 1997 to as high as 65 % in 2005-06, and settling back to 51 % in the most recent year. Electricity savings were higher in every subsequent year following 1997, beginning at just 1 % and ending at 36 %.

Koldus

The savings trends for chilled water, hot water, and electricity consumption for the Koldus Building are shown in bar graph form in Fig. 3.22.

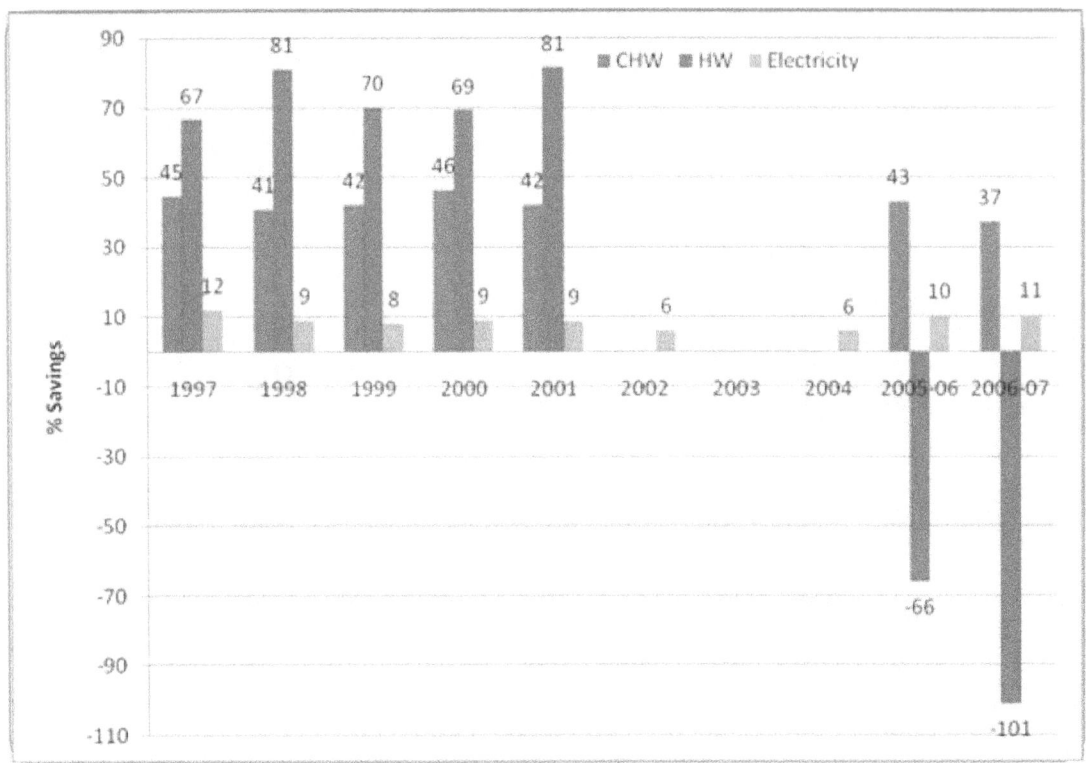

Figure 3.22 Normalized energy savings patterns for the Koldus Building.

The Koldus building demonstrated high levels of persistence in both chilled water and electricity savings over a ten year period. However, it also experienced a huge increase in hot water consumption in the most recent years, even doubling pre-retro-commissioning consumption levels. This was by far the most significant example of savings degradation noted in the ten buildings during the period studied. However, as will be discussed later, this may have been due to metering issues and not actual degradation.

Richardson Petroleum

The savings trends for chilled water, hot water, and electricity consumption for the Richardson Petroleum Building are shown in bar graph form in Fig. 3.23.

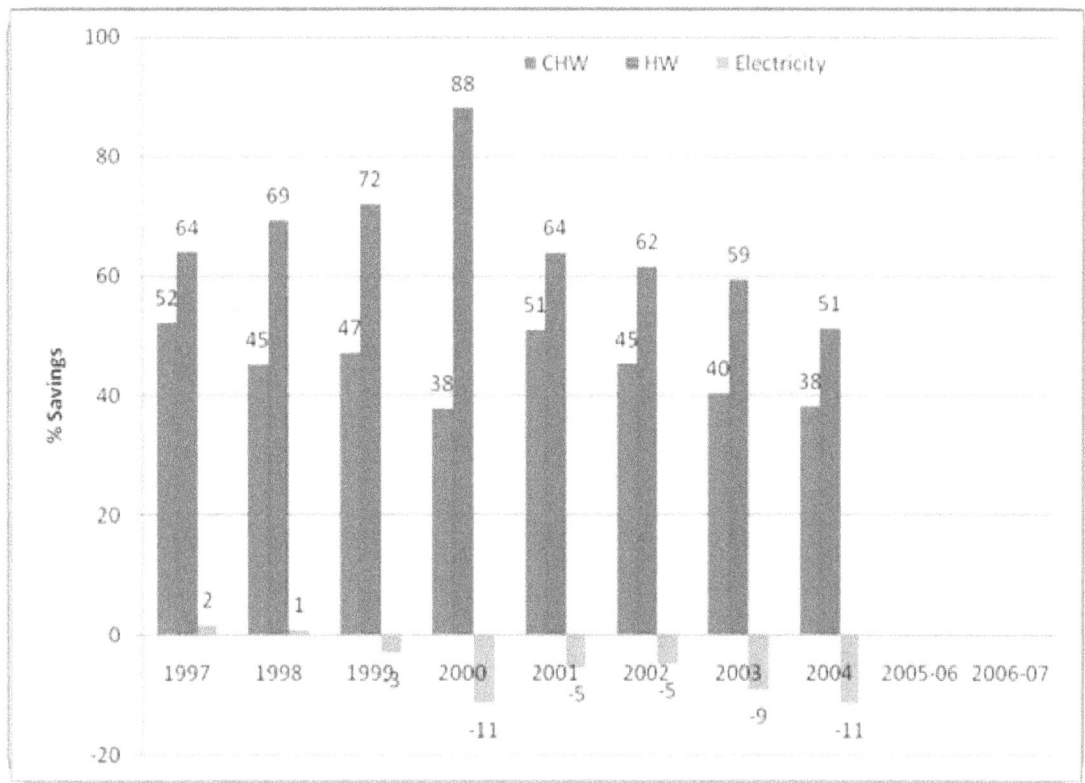

Figure 3.23 Normalized energy savings patterns for the Richardson Petroleum Building.

The chilled water savings for the Richardson Petroleum Building remained fairly steady, and in an eight year period fell only from 52 % to 38 %. The hot water savings increased in each of the first four years after retro-commissioning, peaking at 88 %, but then fell in succeeding years to a level of 51 % in 2004. Electricity savings had fallen to the negative range by the third year after commissioning, and ended in 2004 at -11 %, from the 2 % level in 1997.

Veterinary Medical Center Addition

The savings trends for chilled water, hot water, and electricity consumption for the Veterinary Medical Center Addition are shown in bar graph form in Fig. 3.24.

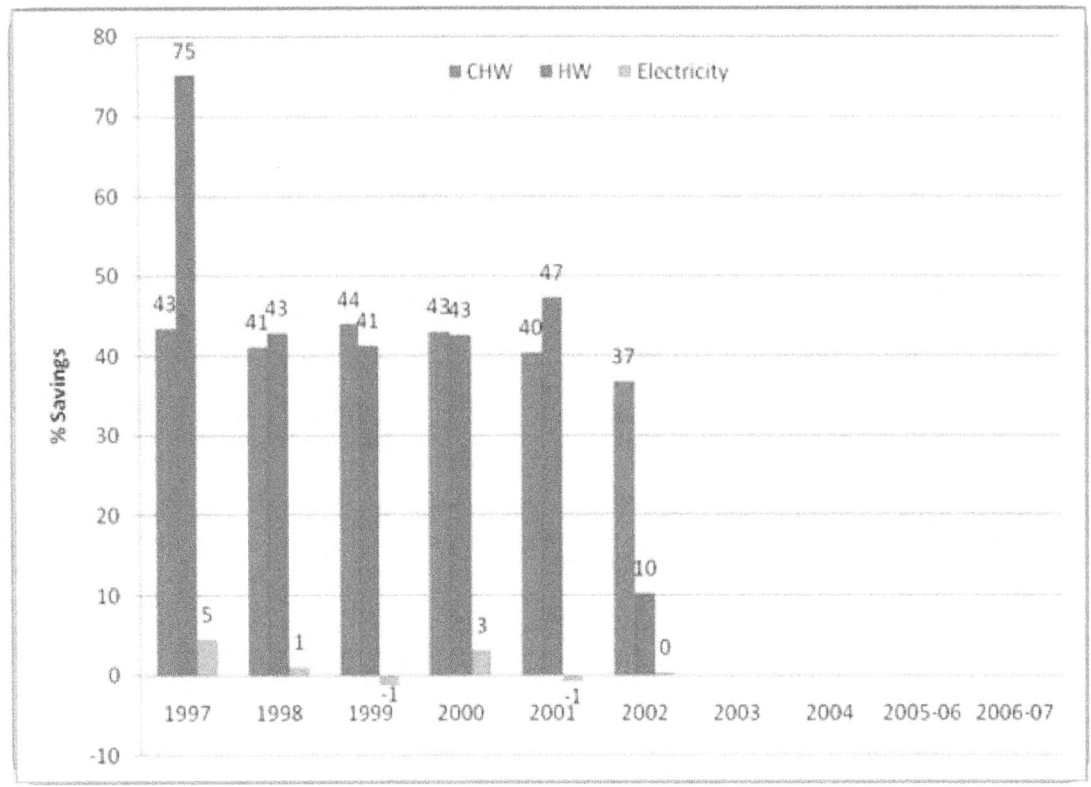

Figure 3.24 Normalized energy savings patterns for the Veterinary Medical Center Addition.

The Veterinary Medical Center Addition had the least amount of reliable energy data available, but a six year period following retro-commissioning was able to be examined. During this time chilled water savings remained consistent, falling only to 37 % in 2002 from 43 % in 1997. Electricity savings essentially degraded to none after a 5 % level initially. Hot water savings was 75 % in 1997, fell sharply to 43 % in 1998, remained very close to that level for the next three years, then fell sharply again to just 10 % in 2002.

Wehner

The savings trends for chilled water, hot water, and electricity consumption for the Wehner Building are shown in bar graph form in Fig. 3.25.

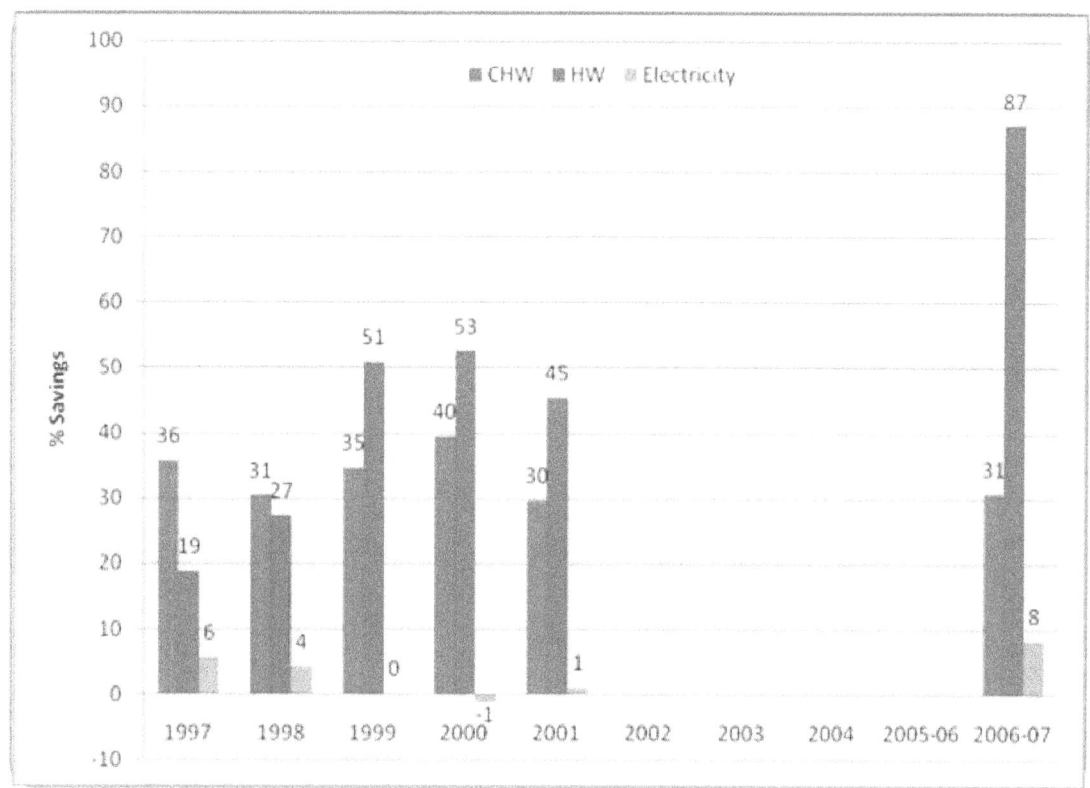

Figure 3.25 Normalized energy savings patterns for the Wehner Building.

The Wehner Building experienced good persistence in chilled water savings over time. Hot water savings increased in the years following retro-commissioning, and remained high in the most recent data year, based on the calibrated simulation model used. Electricity savings degraded some in the years following commissioning, but increased back to a level slightly higher than the 1997 level in the most recent data year, based on the calibrated simulation model.

Zachry Engineering Center

The savings trends for chilled water, hot water, and electricity consumption for the Zachry Engineering Center are shown in bar graph form in Fig. 3.26.

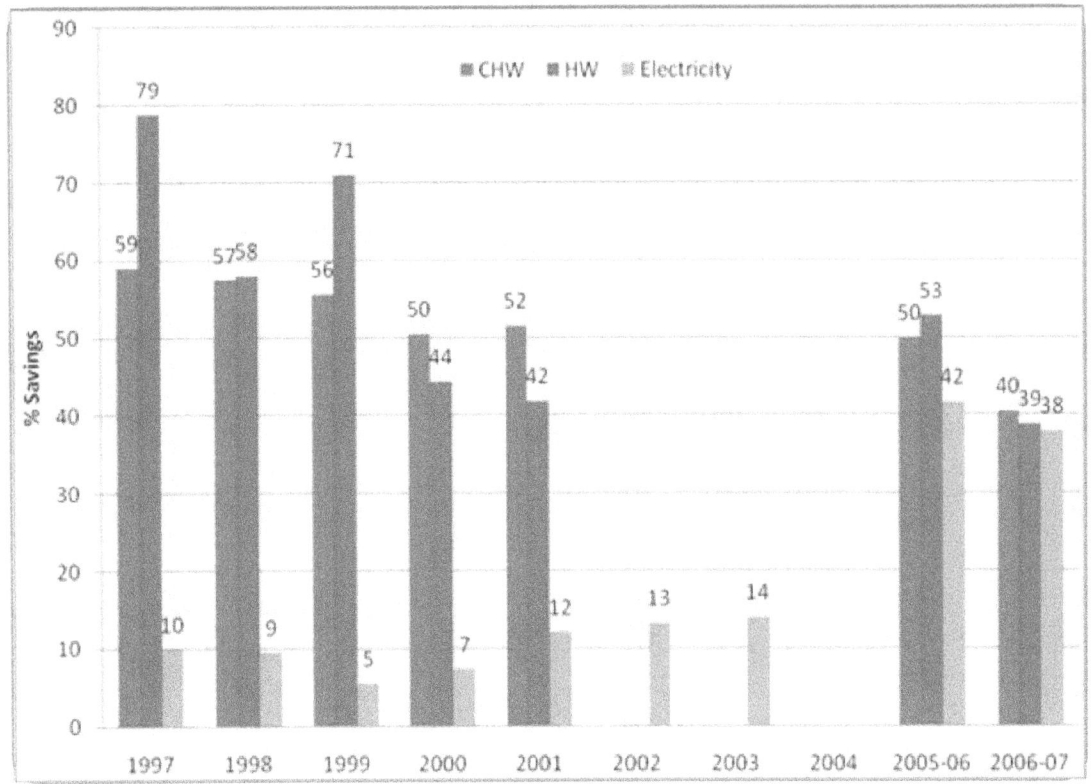

Figure 3.26 Normalized energy savings patterns for the Zachry Engineering Center.

The chilled water savings in the Zachry Engineering Center degraded from 59 % in 1997 to 40 % in 2006-07. The hot water savings fluctuated in the first few years after retro--commissioning. In 1997 it was at its highest level, 79 %, but had degraded to 39 % by 2006-07. Electricity consumption was a different story, however, beginning at 10 % in 1997, holding fairly constant for several years thereafter, then jumping to 42 % in 2005-06, and ending at 38 % in the most recent year.

3.4.3 Summary of Recent Persistence Studies

As a group, the four Japanese buildings studied are consistent with the commissioned buildings that have been examined in the literature review – most continue to show savings over periods that in this case run from 10 to 20 years, while generally showing some decrease in savings over time.

The fundamental conclusion from the additional study of ten buildings at Texas A&M is that cooling and heating savings did not change appreciably during the up to seven years of additional data following the year 2000. The average cooling savings were 45 % in 1997, 34 % in 2000 and 36 % in the last year of good data for each building (average of 2005.9). Heating savings averaged 63 % in 1997, 56 % in 2000 and 55 % in the last year of available data (average 2005-06). It should be noted that follow-up commissioning was conducted in most of these buildings at least once.

3.5 References

ASHRAE. 1996. *ASHRAE Guideline 1-1996: The HVAC Commissioning Process*. Atlanta: American Society of Heating, Ventilating, and Air Conditioning Engineers, Inc.

Bourassa, N.J., M.A. Piette, N. Motegi., "Evaluation of Persistence of Savings from SMUD Retro-commissioning Program – Final Report," Lawrence Berkeley National Laboratory, May 2004. LBNL-54984

Chen, H., S. Deng, and H. Bruner. 2004. *Continuous Commissioning^{SM} Report for Civil Engineering & Texas Transportation Institute (CE/TTI) Building*. Energy Systems Laboratory, Texas A&M University, College Station, TX.

Chen, Hui, Song Deng, Homer Bruner, David Claridge and W.D. Turner, "Continuous Commissioning^{SM} Results Verification And Follow-Up For An Institutional Building – A Case Study," *Proc. 13th Symposium on Improving Building Systems in Hot and Humid Climates*, May 20-23, 2002, Houston, TX, pp. 87-95.

Cho, Sool Yeon, "The Persistence of Savings Obtained from Commissioning of Existing Buildings," M.S. Thesis, Mechanical Engineering Department, Texas A&M University, ESL-TR-02-05-01, May, 2002, 347 pp.

Claridge, D. E., N. Bensouda, S. Lee, G. Wei, K. Heinemeier, and M. Liu. 2003. *Manual of Procedures for Calibrating Simulations of Building Systems*. Prepared for Lawrence Berkeley National Laboratory, Berkeley, CA.

Claridge, D.E., Turner, W.D., Liu, M., Deng, S., Wei, G., Culp, C., Chen, H. and Cho, S.Y., "Is Commissioning Once Enough?," *Solutions for Energy Security & Facility Management Challenges: Proc. Of the 25th WEEC*, Atlanta, GA, pp. 29-36, Oct. 9-11, 2002.

Claridge, D.E., Turner, W.D., Liu, M., Deng, S., Wei, G., Culp, C., Chen, H., and Cho, S.Y., "Is Commissioning Once Enough?" *Energy Engineering*, Vol. 101, No. 4, 2004, pp. 7-19.

Fels, M. 1986. PRISM: An Introduction. *Energy and Buildings* 9(1 and 2): 5-18.

Frank, M., Friedman, H., Heinemeier, K., Toole, C., Claridge, D., Castro, N., and Haves, P., 2005, "Existing Cost/Benefit and Persistence Methodologies and Data, State of Development of Automated Tools, and Assessment of Needs for Commissioning ZEB," Draft report submitted to U.S. Department of Energy under Contract No. DE-AC02-05CH11231, 78 pp., August.

Friedman, H., Potter, A., Haasl, T., and Claridge, D., "Persistence of Benefits from New Building Commissioning," *Proceedings of the 2002 ACEEE Summer Study on Energy Efficiency in Buildings*, Pacific Grove, CA, Aug. 19-23, 2002, pp. 3.129 – 3.140.

Friedman, H., Potter, A., Haasl, T., Claridge, D and Cho, S., "Persistence of Benefits from New Building Commissioning," *Proc. Of 11th National Conference on Building Commissioning*, Palm Springs, CA, May 20-22, 2003a, 15 pp., CD.

Friedman, H., A. Potter, T. Haasl, and D. Claridge, "Report on Strategies for Improving Persistence of Commissioning Benefits – Final Report," July 2003b, 47 pp. Lawrence Berkeley National Laboratory. http://buildings.lbl.gov/hpcbs/pubs/E5P22T5c-Final.pdf

IPMVP 2001. IPMVP Committee, *International Performance Measurement & Verification Protocol: Concepts and Options for Determining Energy and Water Savings*, Vol. 1, U.S. Dept. of Energy, DOE/GO-102001-1187, 86 pp., January.

IPMVP Technical Committee. 2002. *International Performance Measurement & Verification Protocol Volume 1: Concepts and Options for Determining Energy and Water Savings.* U.S. Dept. of Energy: 86.

Katipamula, S., T.A. Reddy, and D.E. Claridge. 1995. Effect of Time Resolution on Statistical Modeling of Cooling Energy Use in Large Commercial Buildings. *ASHRAE Transactions* 101(2).
Liu, M. 1995. *Manual for AirModel.* Energy Systems Laboratory, Texas A&M University, College Station, TX.

Liu, M., Claridge, D. E. and Turner, W.D., 2002, *Continuous Commissioning* [SM] *Guidebook: Maximizing Building Energy Efficiency and Comfort*, Federal Energy Management Program, U.S. Dept. of Energy, 144 pp., Available at
http://www.eere.energy.gov/femp/operations_maintenance/commissioning_guidebook.cfm

Liu, C., Turner, W.D., Claridge, D., Deng, S. and Bruner, H.L., "Results of CC Follow-Up in the G. Rollie White Building," *Proc. 13*[th] *Symposium on* Improving *Building Systems in Hot and Humid Climates*, May 20-23, 2002, Houston, TX, pp. 96-102.

Mills, E., H. Friedman, T. Powell, N. Bourassa, D. Claridge, T. Haasl, and M. Piette, "The Cost-Effectiveness of Commercial-Buildings Commissioning: A Meta-Analysis of Energy and Non-Energy Impacts in Existing Buildings and New Construction in the United States," December 2004. LBNL-56637.

Mills, E., N. Bourassa, M.A. Piette, H. Friedman, T. Haasl, T. Powell, and D. Claridge. "The Cost-Effectiveness of Commissioning New and Existing Commercial Buildings: Lessons from 224 Buildings," Proceedings of the 2005 National Conference on Building Commissioning, Portland Energy Conservation, Inc., New York, New York, May, 2005.
http://www.peci.org/ncbc/proceedings/2005/19_Piette_NCBC2005.pdf

National Climatic Data Center (NCDC). Viewed 21 September 2006.
http://cdo.ncdc.noaa.gov/pls/plcimprod/cdomain.abbrev2id

Peterson, Janice, "Evaluation of Retro-commissioning Results After Four Years: A Case Study," Proceedings of the 2005 National Conference on Building Commissioning, Portland Energy Conservation, Inc., New York, New York.

Selch, M. and J. Bradford, "Re-commissioning Energy Savings Persistence," Proceedings of the 2005 National Conference on Building Commissioning, Portland Energy Conservation, Inc., New York, New York, May, 2005.

Shao, X. 2005. First Law Energy Balance as a Data Screening Tool. M.S. Thesis, Texas A&M University, Department of Mechanical Engineering, College Station, TX.

Toole, C., "The Persistence of Retrocommissioning Savings in 10 University Buildings," M.S. Thesis, Mechanical Engineering Department, Texas A&M University, Dec, 2009,

Turner, W.D., Claridge, D.E., Deng, S., Cho, S., Liu, M., Hagge, T., Darnell, C., Jr., and Bruner, H., Jr., "Persistence of Savings Obtained from Continuous Commissioning[SM]," *Proc. of 9th National Conference on Building Commissioning*, Cherry Hill, NJ, p. 20-1.1 - 20-1.13, May 9-11, 2001.

Wei, G., M. Liu, and D.E. Claridge. 1998. Signatures of Heating and Cooling Energy Consumption for Typical AHUs. *Proceedings of the Eleventh Symposium on Improving Building Systems in Hot and Humid Climates.* Fort Worth, TX.

Yamaha, Motoi, "Persistence Surveys in Japan," Powerpoint presentation A47-C2-M5- JP-CU-1 at IEA Annex 47 Meeting 5, Kyoto, Japan, October, 2007.

4. TOOLS TO ENHANCE PERSISTENCE

4.1 An Automated Building Commissioning Analysis Tool (ABCAT)

4.1.1 Introduction

In the United States, slightly more than one-third of the total primary energy consumption is used in the building sector. Commercial buildings alone cost 18 % of the total energy use in the U.S. in 2007 (Energy Information Administration. 2007). Energy conservation programs for the building sector would contribute to the reduction of energy sources waste. Building commissioning services, which either ensure that building systems are installed and operated to provide the performance envisioned by the designer or identify and implement optimal operating strategies for buildings as they are currently being used, have proven to be successful in saving building energy consumption. A broad and major study of 224 new and existing commercial buildings in 21 states across the country, commissioned by 18 different commissioning service providers, netted a median savings of 15 % of whole building energy use (Mills et al. 2005). The Energy Systems Laboratory at Texas A&M University (TAMU) started Continuous Commissioning[®1] (CC[®]) in 1996. The CC[®] process has produced average energy savings of about 20 percent without significant capital investment in over 150 large buildings in which it has been implemented (Claridge et al. 2004).

Though commissioning services are effective in reducing building energy consumption, the optimal energy performance obtained by commissioning may subsequently degrade, as described in Chapter 3. The persistence of savings is a significant topic of concern. Claridge et al. (2004) presented the results of a study of the persistence of savings in ten university buildings that averaged an increase of chilled water (CHW) and hot water (HW) costs by 12.1 % over a two year period post-commissioning. Almost 75 % of this increase was caused by significant component failures and/or control changes that did not compromise comfort but caused large changes in consumption. The remainder was due to control changes implemented by the operators (Claridge et al. 2004, Turner et al. 2001). The major increases were not identified until two years had passed, and hundreds of thousands of dollars in excess energy costs had already occurred. Obviously there is a need for a simple, cost efficient automated system that can continuously monitor building energy consumption, alert operations personnel early upon the onset of significant increases in consumption and assist them in identifying the problem. The Automated Building Commissioning Analysis Tool (ABCAT) is one of several such tools for maintaining the optimal energy performance in a building.

ABCAT was originally initiated by Lee and Claridge (2003), and has developed to an advanced prototype by Curtin et al. (2007), along with demonstrating its effectiveness in live and retrospective building implementations. This report describes the functions of the advanced prototype ABCAT tool and provides a summary of its live testing results on six buildings and retrospective testing results on five buildings. The tool has not yet been commercialized.

ABCAT Description

The fault detection and diagnosis approach to be undertaken in ABCAT will be applied to the whole building energy consumption level and is simplified to aid in the practicality of its implementation outside of the university and research lab setting. First, a building energy simulation model using the American Society of Heating, Refrigerating and Air-Conditioning Engineers (ASHRAE) simplified energy analysis procedure (Knebel et al. 1983) is established and calibrated based on the building CHW and HW consumption in the baseline period chosen from a post-commissioning time period when the building's operation is considered to be optimal. Second, subsequent CHW and HW consumption is predicted by the model using future weather data and building electricity consumption. Third, both the simulated and measured consumption are passed to the data analysis routine that generates building performance plots, compares and performs calculations on the simulated and measured consumption data, applies fault detection methods, and reports diagnostic and energy consumption statistics. Finally, the user of the tool evaluates the data presented and determines whether or not there is a fault that requires action. If a fault is identified, the user or other experts can use the diagnostic information provided by ABCAT to help identify and correct the fault, and follow up observations should observe a return to expected performance.

The ABCAT is initially setup in a building through the following sequence of steps:

<u>1. Define a Baseline Consumption Period and Collect Baseline Measurements</u>
The baseline period should correspond to a time when the building mechanical systems are known to be operating correctly, typically post new building commissioning (Cx) or existing building commissioning (EBCx). The length of baseline can be a minimum of four weeks if during the swing seasons where a wide range of outside air temperatures is experience and heating and cooling systems are both operating. Required measurements include whole building heating (WBHeat) whole building cooling (WBCool), whole building electric (WBElec), ambient outside air temperature and relative humidity or dew point temperature, all recorded in hourly intervals.

Figure 4.1 describes the consumption monitoring that is required for the ABCAT. Ideally the WBHeat and the WBCool would be obtained by Btu metering of chilled and hot water, but these values could also be obtained by modeling the chiller and boiler if interval meters exist that monitor chiller electric loads and natural gas consumption.

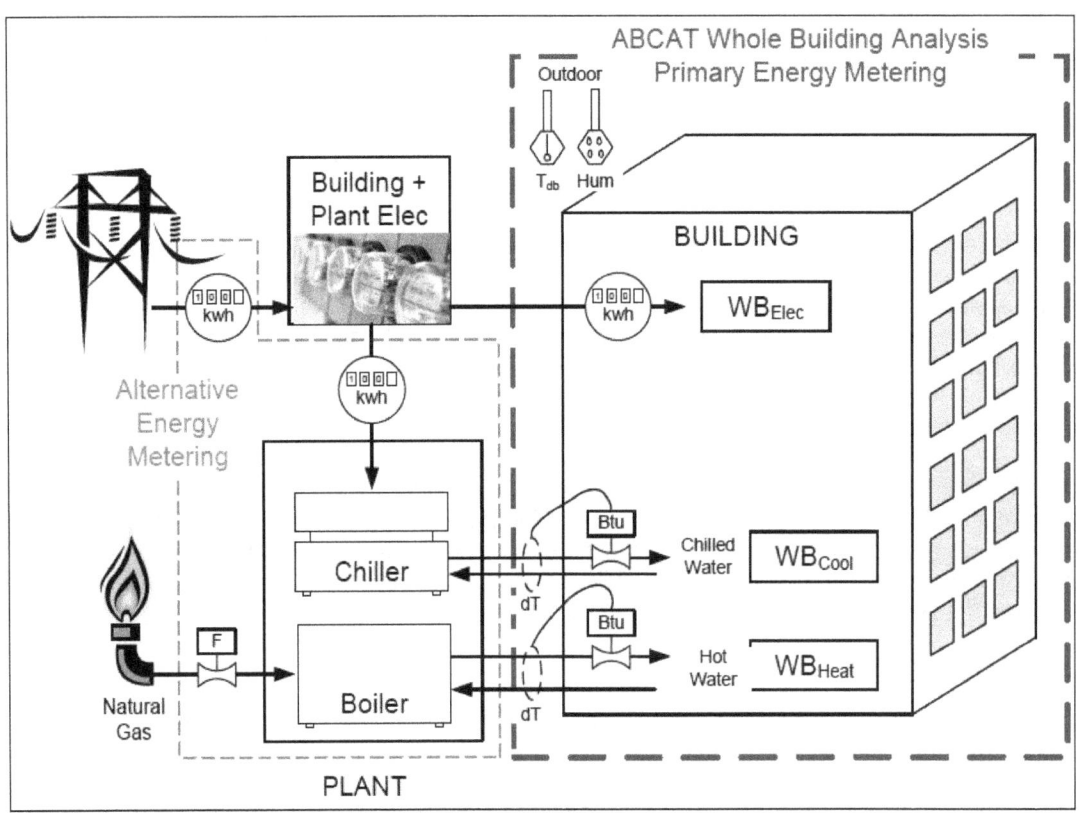

Figure 4.1. Consumption metering requirements for ABCAT

2. Obtain Building and Air Handling Unit System Details

The key characteristics of the building and its heating ventilation and air-conditioning (HVAC) systems that must be included in the model are:

(1) Envelope Area and Heat Transfer Coefficient

(2) Solar Radiation Load

(3) Internal Heat Gain from Equipment and Lighting (taken as fraction of measured electric load)

(4) Deck Temperature Schedules

(5) Maximum and Minimum Air Flow Rates,

(6) Outside Air Intake and Economizer Settings

(7) Occupancy Schedule

(8) AHU Operation Schedule

(9) Humidification Operations.

Liu et al. (1998) describes the steps for an initial value selection of these parameters in a model with similar input requirements as the ABCAT.

3. Establish Initial Values of Inputs for the Simulation Model and Calibrate the Model

Generate an input file for simulation based on measured data and system information, and calibrate or tune the model inputs until desired accuracy is achieved.

4. Correct for Bias in Model

Provide a final adjustment to simulation model by calculating the mean bias error (MBE) and subtracting this amount from to the model so that the MBE of the model is zero for the baseline period. Even a small systematic bias in the simulation will decrease the sensitivity of the fault detection process.

5. Program Regular Data Transfer to ABCAT

Develop a method by which the required measured inputs can regularly be updated and passed to the ABCAT program. In the current test facilities, Visual Basic for Applications programs link the ABCAT with consumption data files, sorts, fills missing data with linear interpolation when applicable, summarizes and imports the data into the ABCAT program in its required format.

Once the ABCAT is configured for the particular building through steps 1 to 4, the program is ready for execution.

Figure 4.2 is a process flow diagram which visually describes the following five steps to the ABCAT methodology:

1. Import Measured Data

Evoke the program developed in step 5 of the initial setup steps 1 to 5 from the ABCAT program.

2. Simulate CHW and HW Consumption

The required inputs are passed to the energy simulation routine, where the CHW and HW consumption is simulated. A detailed flow diagram of the ABCAT process through to the execution of the simulation is summarized in Appendix A.

3. Data Analysis

The simulated consumption and measured consumption are passed to the data analysis routine that generates the building performance plots, compares and performs calculations on the two values, applies fault detection methods, and reports diagnostic and energy consumption statistics. The current graphical layout of the ABCAT is detailed in Appendix B.

4. Evaluation

The user of the tool is to evaluate the data presented and arrive at the conclusion as to whether or not a fault of enough significance exists such that action is required. The user plays an important role in defining fault triggers and manipulating the plotted data with easily adjustable parameters to suit their site specific preferences. One of the primary metrics established to aid in the user decision is the "Cumulative Energy (or Cost) Difference" plot, previously used by Haberl and Vajda (1988), which accumulates the energy residuals of persistent deviations from measured consumption by adding them to that of the total of the previous day (multiplies the accumulated energy by a user specified utility cost for the Cost Difference plot). The cost plot presents the deviations in the universally understood language of dollars and cents, which is expected to help compel users of the ABCAT to act in the case of a fault.

5.Action

If action is deemed necessary, the type of action taken will depend upon if the faulty condition observed is determined to be a result of a required change in operations (where the simulation model would have to be recalibrated) or if it was caused by a system or component failure or a change in control to a less than optimal setting (where repair, maintenance or a control change may be in order). If a fault is identified, the user or other experts can use the diagnostic information provided by ABCAT to help identify and correct the fault, and follow up observations should observe a return to expected performance.

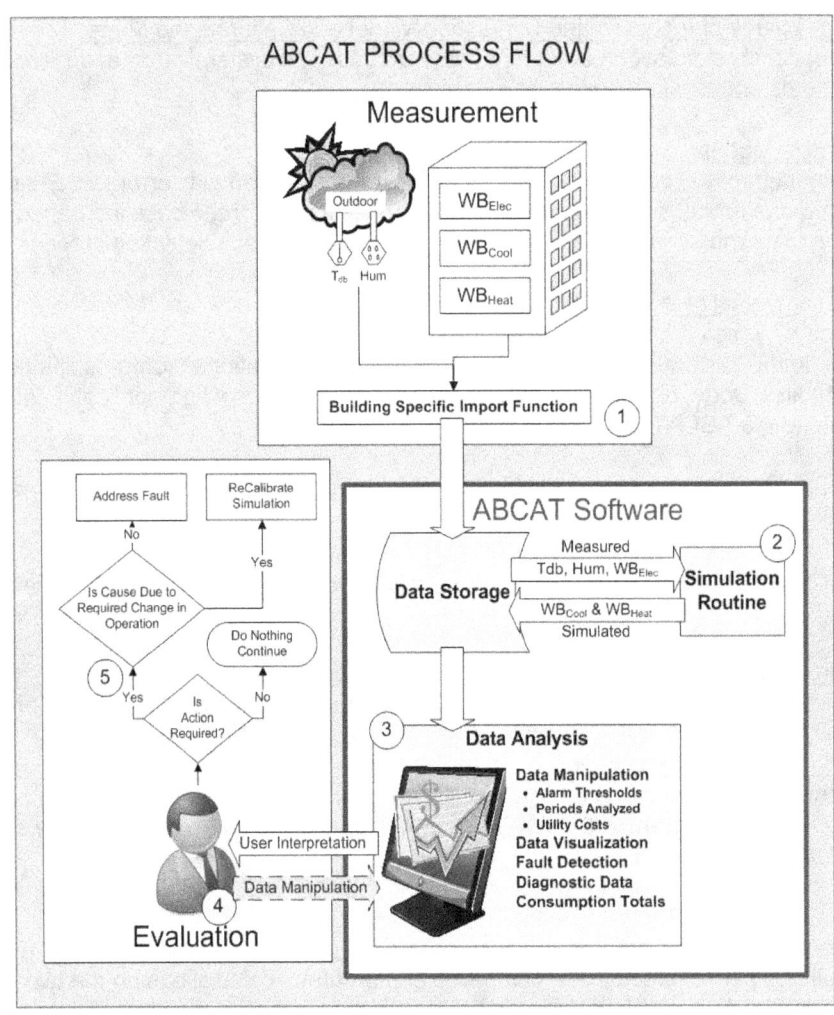

Figure 4.2 ABCAT flow diagram

Project Results

Live Test Cases

The ABCAT was implemented in six live building situations, with various levels of automation and file manipulation were built into a specific data collection process for each building based on its unique conditions regarding data availability and format. The testing of the ABCAT in the six buildings provided a live learning scenario that helped to influence continued developments, and a summary of these test cases is provided in Table 4.1.

Table 4.1 Test buildings, results and findings from live ABCAT implementation

Building Description	Location	Test Period	Results and Findings
7618 m² 82,000 ft² university dining facility	College Station, Texas	Mar 2005 – July 2007	• Detected excess cooling energy fault related to excessive latent cooling from low discharge air temperature on 2 of 3 Outside Air Handling Units – Summer 2006 shown in Fig. 4.3.
44779 m² 482,000 ft² computing services facility	Austin, Texas	May 2005 – July 2007	• Detected significant decrease in measured cooling energy due to meter calibration – Oct 2005 (Figure 4.4). • A second fault, significant excess cooling energy was detected in Nov 2006 Fig. 4.5). • Also demonstration of successful short-term adaptation of simulation to multiple baseline changes.
16723 m² 180,000 ft² office building	Albany, New York	Jan 2007 – July 2007	• Successful monitoring of heating energy savings following implementation of EBCx measures Fig. 4.6). • Training and support for two ABCAT testers.
16723 m² 190,000 ft² high-rise office building	Omaha, Nebraska	Feb 2007 – July 2007	• Confirmation of optimal heating and cooling energy through continued tracking. • Identification of HW metering failure (Figure 4.7).
12356 m² 133,000ft² university teaching building	College Station, Texas	June 2007- May 2009	• Confirmed excess outside air increased CHW use from 06/2008-08/2008 • CHW increase from 9/15/2008-12/05/2008 and 02/11/2009-05/18/2009 observed.
6625 m² 67,000ft² university office building	College Station, Texas	Nov 2007- May 2009	• Identified apparent CHW meter problem during 11/2008. • Noted apparent HW meter recalibration during 2008.

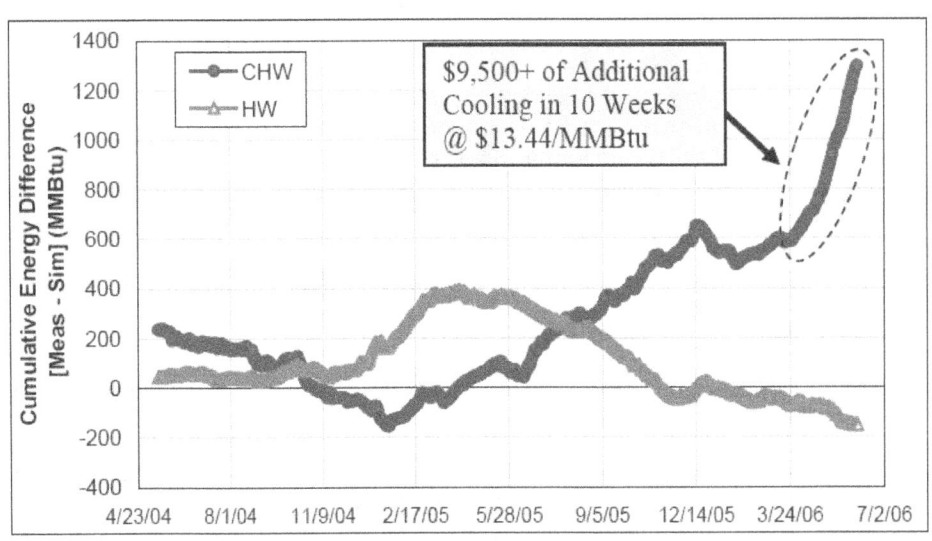

Figure 4.3 Sbisa Dining Hall cumulative energy difference meas – sim (MMBtu) with simulation calibrated to period of 5/01/2004 to 06/27/2006

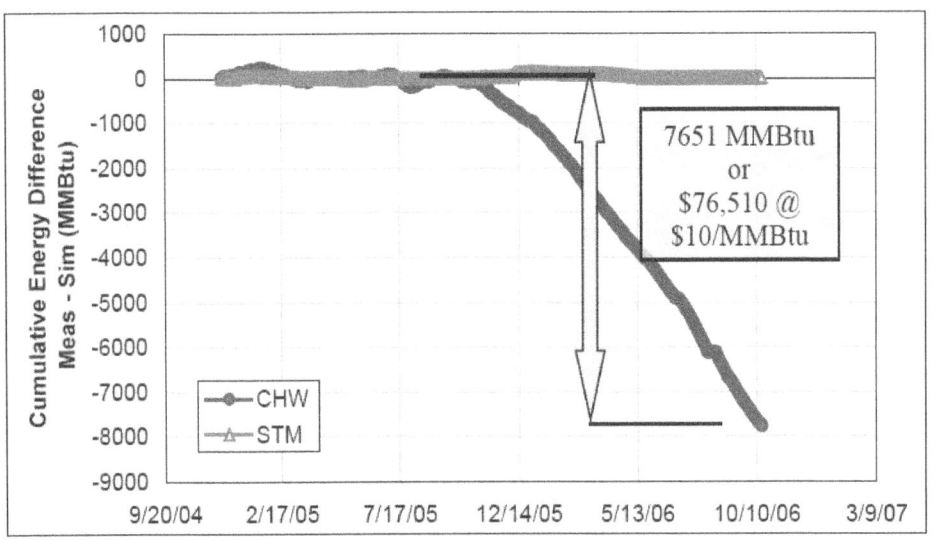

Figure 4.4 Computing Services Facility ABCAT cumulative energy difference meas – sim (MMBtu) with simulation calibrated to period of 12/01/2004 to 10/27/2005

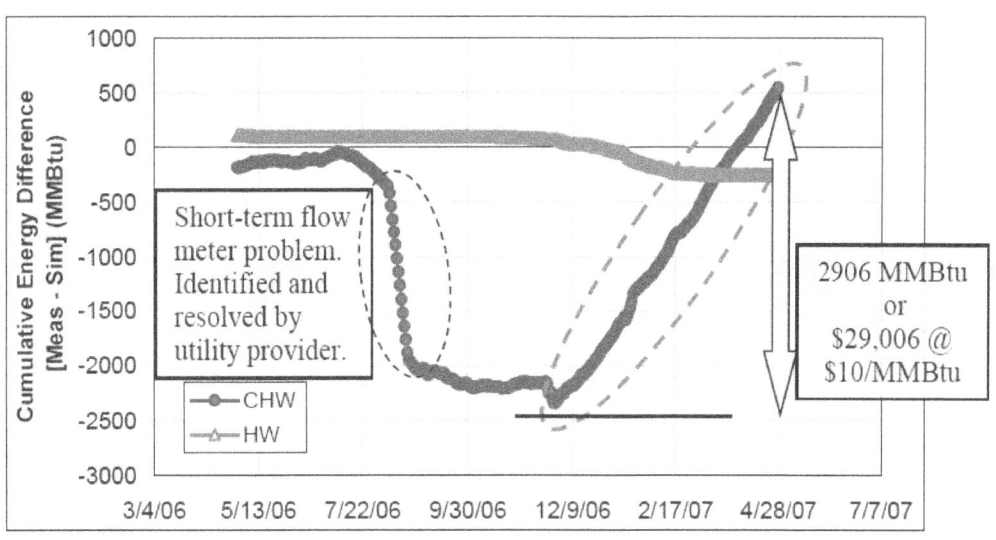

Figure 4.5 Computing Services Facility cumulative energy difference for period starting 04/29/2006 for 1 year after simulation recalibrated to period of 10/27/2005 – 5/19/2006

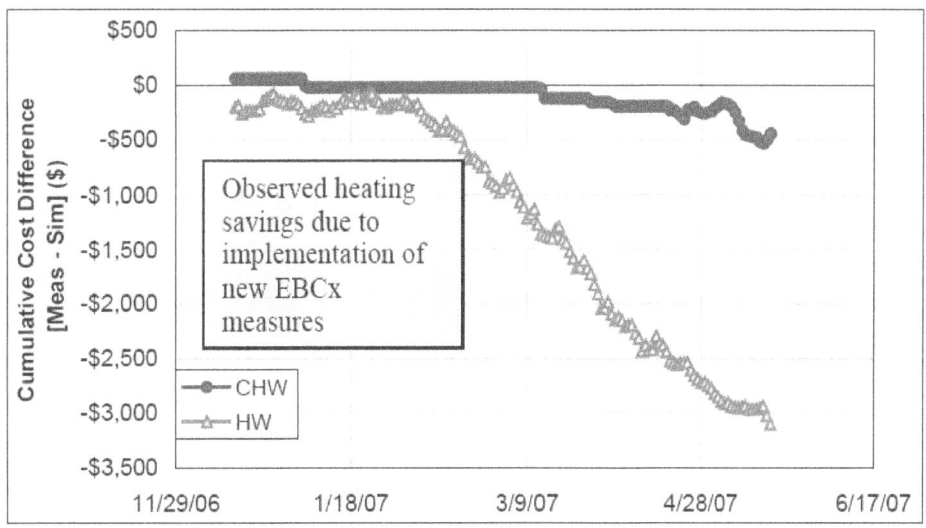

Figure 4.6 DASNY cumulative cost difference ($15/MMBtu heating, $10/MMBtu cooling)

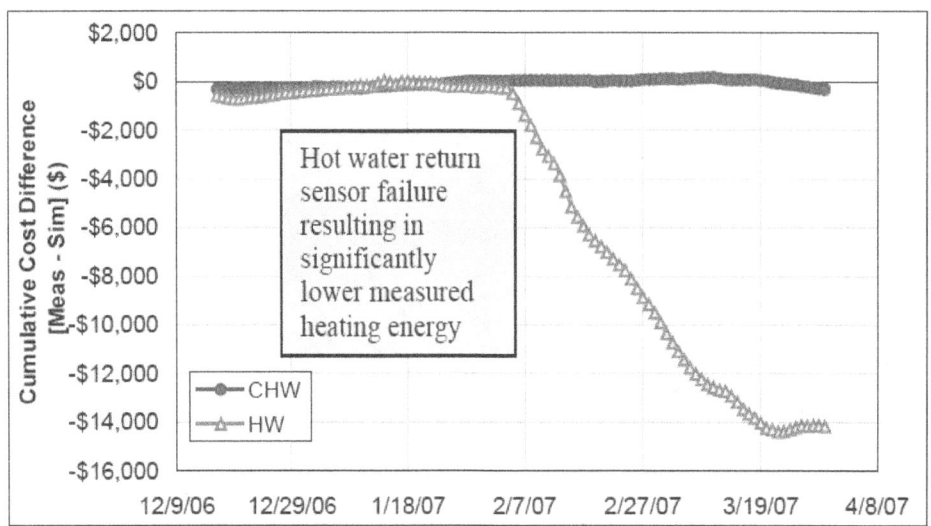

Figure 4.7 OPPD energy plaza cumulative cost difference
($15/MMBtu heating, $10/MMBtu cooling)

The testing of the ABCAT in these six buildings, the identification of the faults and the diagnostic reasoning that followed, helped shaped some of the specific ideas as to the developmental direction of the ABCAT. Some of the keys points to take away from these test experiences are the following:

- Whole building analysis can provide valuable diagnostic information
- Accumulated deviations from optimal performance provide the good indicator of significant faults that persist, and cost information
- The value of ABCAT does not appear to lay in daily short-term observations, but rather observations on the order of weeks to months.
- The advantage of using a first principles simulation model can be seen with occasional recalibrating requirements due to changes in building operations

Retrospective Test Cases

In order to further test the capabilities of ABCAT, a multiple building retrospective test is performed. Five buildings on the Texas A&M University campus which had previously been studied in a commissioning persistence study (for the years of 1996 to 2000), had fairly complete consumption data sets, historical documentation as to commissioning measures implemented, and documentation of some control system set point changes during the period analyzed. It was expected that an analysis with ABCAT of a span of more than 15 building years, would provide some immediate feedback into the fault detection and diagnostic capability of the tool.

The "Cumulative Energy (or Cost) Difference" plot can visually detect a fault and show how the fault influences energy cost. Because visual fault detection depends heavily on personal subjective experience, the "Days Exceeding Threshold" plot was developed and added into ABCAT to detect faults analytically. It is drawn based on the simple standard that identifies a fault if the deviation between the measured and simulated consumption is greater than one standard deviation in the baseline period and persists for at least 30 d. The reason for choosing 30 d as the fault definition is that the typical utility meter reading interval is one month. Every point in the plot represents the number of days in the next 30 d (including the day on which the point is plotted) where consumption has been at least one standard deviation above or below expected consumption. For example, a point at ±10 means there are 10 d of the next 30 d when the measured consumption is more than one standard deviation above or below the simulated consumption. Thus a fault period appears as one or more points at ±30 on the plot. Compared with the "Cumulative Energy (or Cost) Difference" plot, the "Days Exceeding Threshold" plot permits relatively precise identification of the time that a fault starts or ends and provides more objective fault detection metrics. In the retrospective cases, the "Days Exceeding Threshold" plot is used as the chief fault detection criterion.

Eighteen faults were detected in 15 building-years of consumption data with the "Days Exceeding Threshold" plot. One of the eight detected CHW faults and six of the ten detected HW faults are verified by the historical information. The remaining fault diagnoses remain unconfirmed due to data quality issues and incomplete information on maintenance performed in the buildings. A summary of these test cases is provided in Table 4.2.

Table 4.2 Building Faults Detected in five buildings

Building Description	Test Period	Results and Findings
192,000 ft^2 university teaching building	Jan 1997 – Dec 2000	• Detected two excess heating energy faults (HW Fault #1 and 2 in Fig. 4.8) which might be related to scaling problems on the HW meter. • Detected one decrease in measured cooling energy (CHW Fault in Fig. 4.8) which might be caused by an increase in the cold deck temperature.
165,000 ft^2 university teaching building	Nov 1996 – Dec 2000	• Detected significant decrease in measured heating energy (HW Faults #1 to #4, and #6 in Fig. 4.9) due to a HW meter problem. • Detected one excess heating energy fault (HW Fault # 5 in Fig. 4.9) due to the problems the Kleberg Center experienced after April 1999 as documented in Chen et al (2002) • Detected five excess cooling energy faults (CHW Faults #1 to #5 in Fig. 4.9). CHW Fault # 1 to #3 and #5 can't be diagnosed because of the data quality issues. The reasons for CHW Fault #4 were the same as for HW Fault # 5.
180,000 ft^2 university teaching building	Mar 1997 – Dec 2000	• Detected a significant decrease in measured heating energy (HW Fault # 1 in Fig. 4.10) which may be related to a HW meter problem. • Detected one excess heating energy fault (HW Fault # 2 in Fig. 4.10) which may be related to an increase in minimum airflow ratio and hot deck temperature. • The "Cumulative Cost Difference" plot (Figure 4.11) shows that the CHW consumption deviation over four years The maximum CHW consumption deviation over four years is approximately 1 % of the cumulative consumption. This indicates that the simulation is capable of accurately predicting consumption if there are no significant changes in the building.
115,000 ft^2 university teaching building	Jan 1998 – Dec 2000	• Neither a CHW fault nor a HW fault was detected on the "Days Exceeding Threshold" plot (Figure 4.12).
131,000 ft^2 university teaching building	Aug 1996 – Dec 2000	• Detected two excess cooling energy faults (CHW Fault # 1 and #2 in Fig. 4.13) which can't be diagnosed because of data quality issues.

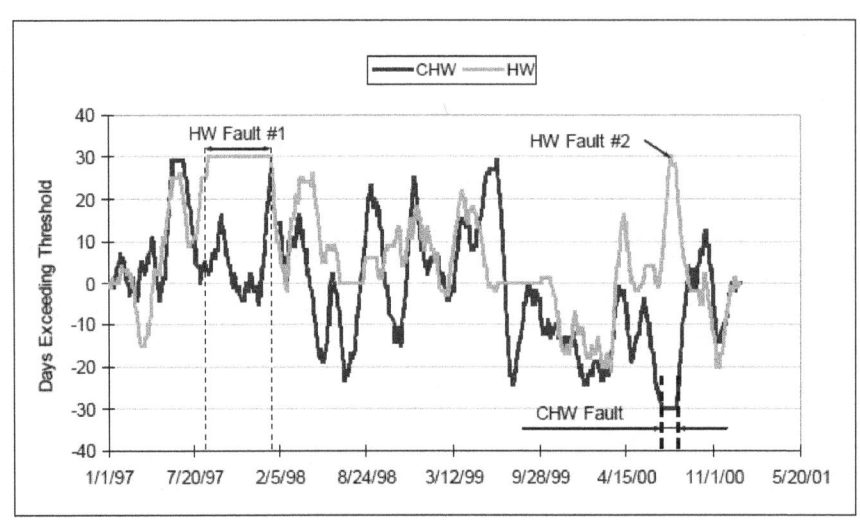

Figure 4.8 Days exceeding threshold in 30 d periods
from 01/01/1997 to 12/31/2000 for the Wehner Building

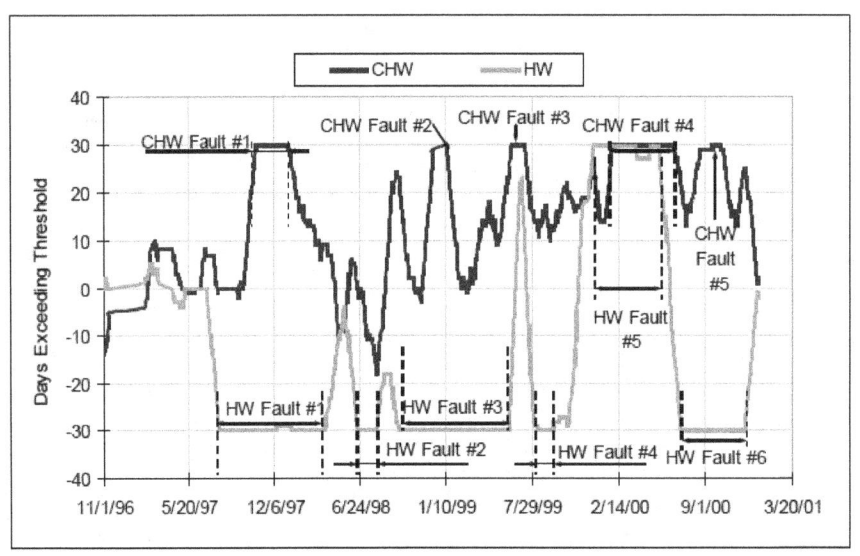

Figure 4.9 Days exceeding threshold in 30 d periods
from 11/01/1996 to 12/31/2000 for the Kleberg Center

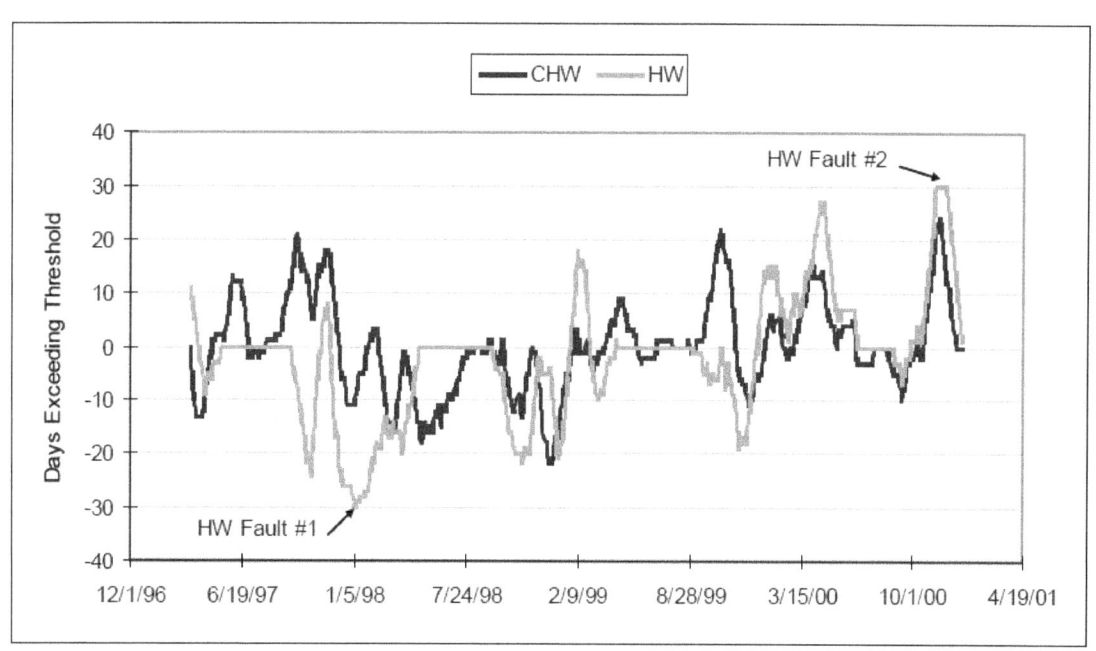

Figure 4.10 Days exceeding threshold in 30 d periods
from 03/19/1997 to 12/31/2000 for the Eller O&M Building

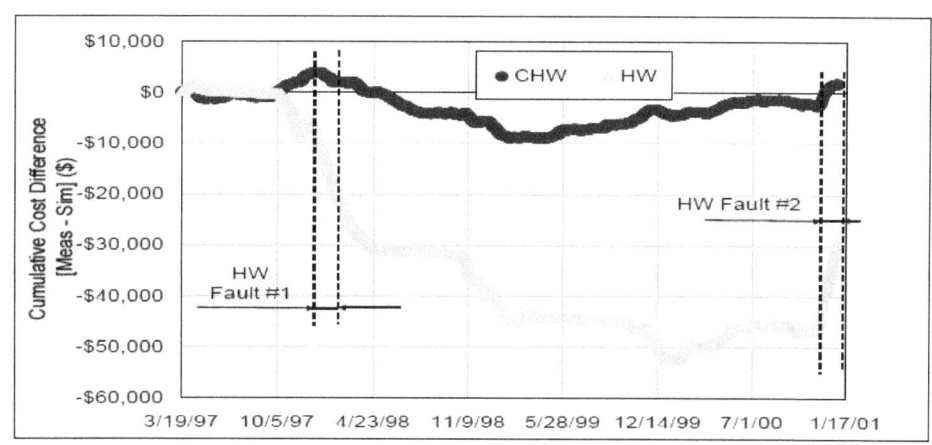

Figure 4.11 Cumulative heating and cooling cost differences for the period
of 03/19/1997 to 12/31/2000 for the Eller O&M Building
(Assuming $10 and $15/MMBtu for CHW and HW respectively)

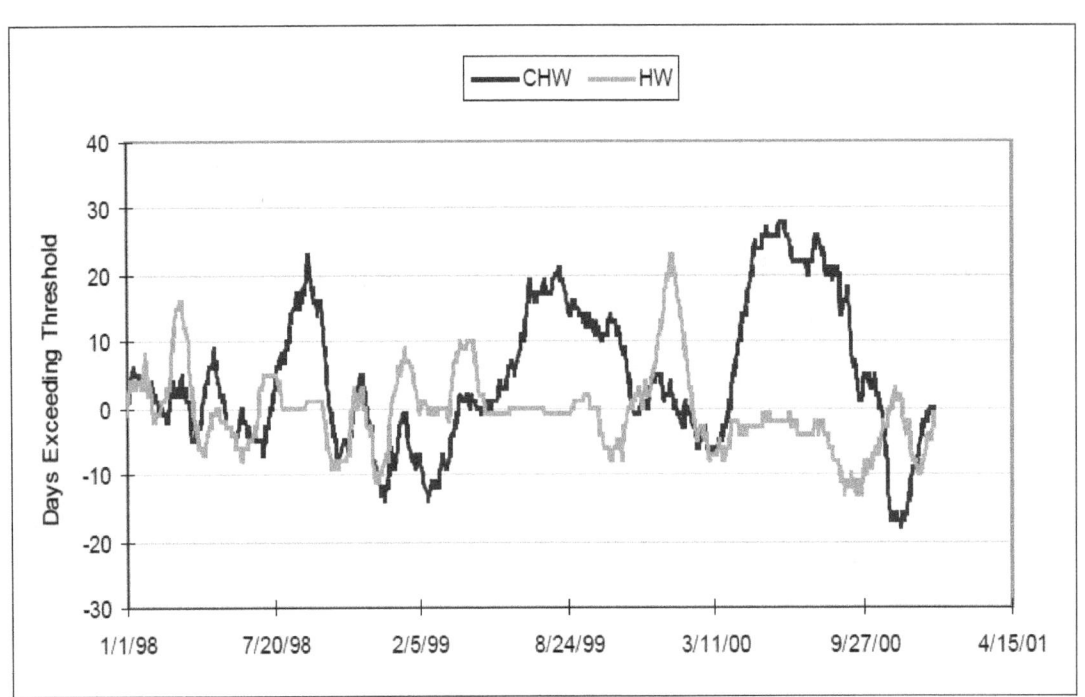

Figure 4.12 Days exceeding threshold in 30 d periods
from 01/01/1998 to 12/31/2000 for the Veterinary Research Building

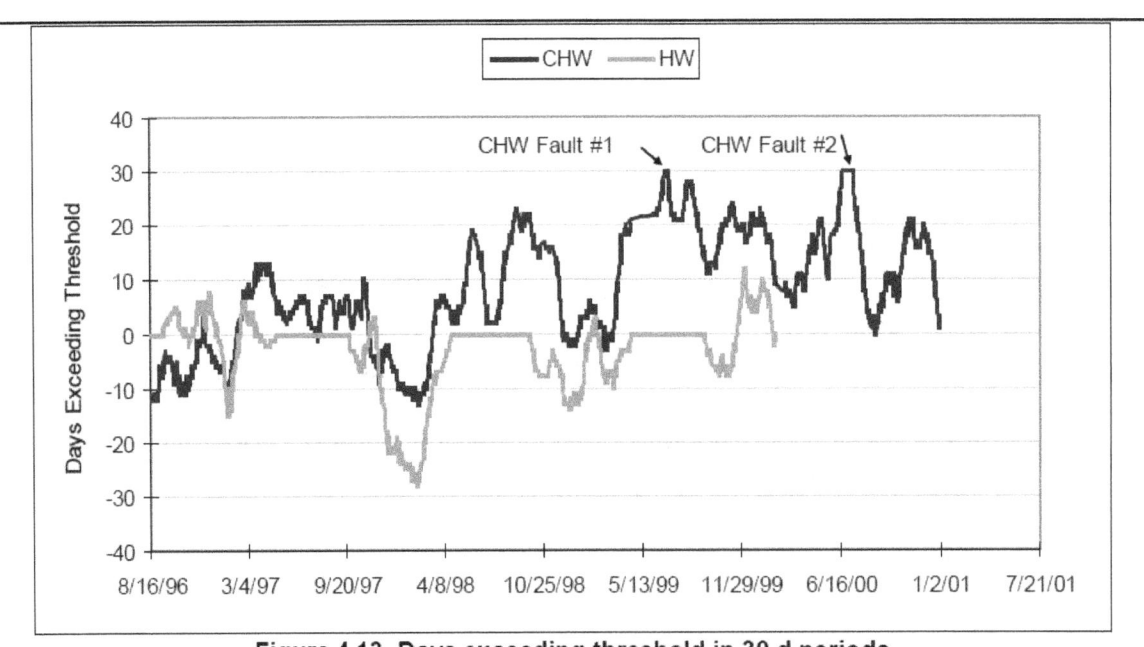

Figure 4.13 Days exceeding threshold in 30 d periods
from 08/16/1996 to 12/31/2000 for Harrington Tower

The retrospective test cases provided an opportunity to test the simulation capabilities of the ABCAT in five additional buildings of varying types and functions, and indicate ABCAT is a promising fault detection and diagnosis tool for post-commissioning use in buildings.

ABCAT Layout

Interface

The ABCAT is laid out as any typical Microsoft Excel file, with multiple worksheets and chart sheets accessible by the colored tabs at the bottom of the screen. The Interface sheet (Fig. 4.14) is the gateway of communication between the user and the tool, and includes the following:

- The dates of the periods analyzed can be adjusted
- Various alarm thresholds can be modified to user preferred levels
- Utility cost information can be specified
- Folder and file locations can be setup for importing and saving data files
- The calibrated simulation statistical results for the baseline
- Consumption totals and diagnostic summary of the period analyzed

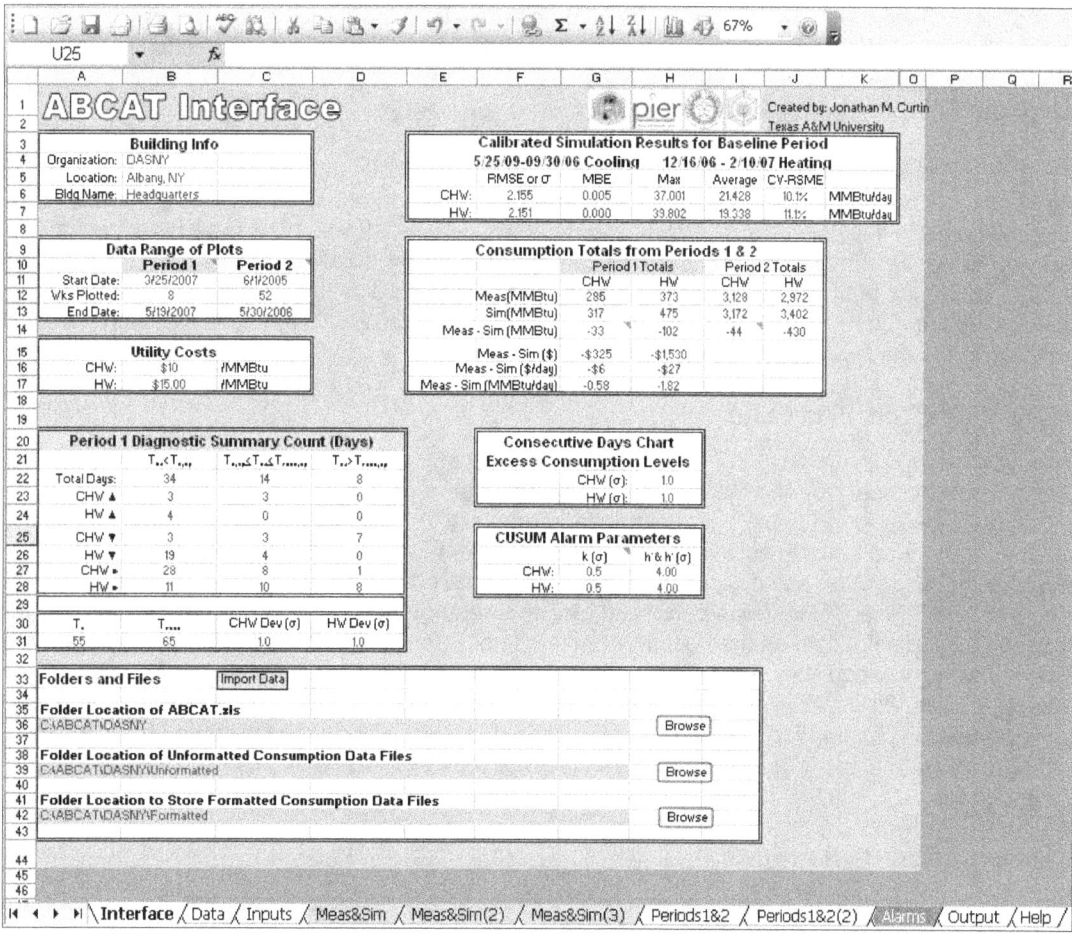

Figure 4.14 The ABCAT user interface

Other Features

- Multiple plots on each of five chart sheets providing performance data comparing measured and simulated or two periods of measured data.
- Quick day-type and date association for all plotted points with double point click
- Scroll through time with the scroll bars

- Daily data summarized and stored in the tool such that the simulation can run for any period without user concern of reprocessing or collecting required inputs.

Analysis of Comments and Feedback

In the developmental process of the ABCAT, informal comments and feedback from users, potential users, as well as management and advisory committee members involved in the project were used to guide many choices about the features implemented in the advanced prototype tool. Although no statistical significance can be assigned to this feedback due to the small number of participants involved and no formal collection methodology was used, this feedback has nonetheless been beneficial in implementing upgrades to the tool, and setting the course for future developmental and testing steps.

General Approach

A common theme that has resonated almost universally through all those involved in testing or evaluating the ABCAT is that there is interest and a market for *simple* energy monitoring and fault detection and diagnostic tools. Over-complication of a tool can immediately lead to reservations from any potential future user that is not about ready to quit his day job to learn how to operate and continually manage the tool. Participants expressed general support for an adequate modeling system that could improve upon the setup complications of programs like DOE-2 or Energy-Plus.

Several positive remarks came with the presentation of the cumulative cost difference chart in the ABCAT. Thoughts were that this would provide a motive for users to act, or at least help the user in providing justification if higher level approval would first be required. It was also stated that detection, and clearly expressing the significance of faults carries a greater weight than the diagnostics, since experts will be able to find the fault if called upon, but if they don't know it exists they might not ever move to address it. One comment that is fitting for the future development is *"Do you put the expert in the tool or leave the expert in the field?"*

Summary of New York Pilot Feedback

The protoype tool was installed in the headquarters building of the Dormitory Authority of New York and used by the building operator and the commissioning engineers working on the building for several months. The tool was found to be "very helpful and beneficial for tracking energy consumption on a higher level" and "good for both building owners and operators", although it was stated that the tool "cannot take the place of on-site diagnostics". On weekly time intervals the required consumption data was imported into the tool, and it was perceived by one user that the optimal time interval for using the tool is weekly. Microsoft Excel as the host program was considered "Good" as far as file size, speed of execution, graphical capabilities, data storage general file layout, familiarity and ease of operation were concerned. A preference was expressed for greater clarity with labels on the Interface sheet for user manipulated fields, consumption period totals for both defined periods on the Interface sheet, and a linking to greater granularity (hourly) than daily data. One user expressed an interest in a lesson to calibrate the tool.

Software Layout and Performance

With the use of Microsoft Excel as a host to the ABCAT, speed of execution, program flow and the size of the program were concerns, although current performance capabilities were viewed as favorable. Recommendations for linking to Microsoft Access were made, which could strengthen data storage capabilities, and allow for storage of smaller time interval and supporting data that would not be feasible to manage within Excel alone. The familiarity of most users with the general function of Excel was seen as a bonus.

As far as the graphical presentation of the tool was concerned, positive feedback was received from the multiple plots per chart sheet layout, scroll bars for zooming, and pop-up window feature for identifying day type and date of specific data points. The Interface sheet of the tool was upgraded in response to recommendations for including data summary tables, and ease of identifying user control options.

Additional recommendations of including day typing, highlighting the most recent data on plots, and general "cleaning" of the plot areas were found to be valuable. Due to the variety of viewing preferences by user, options are provided in the ABCAT for rearranging the existing or creating new plots

Conclusions

ABCAT is a simple, cost efficient automated tool for maintaining the optimal energy performance in a building. It can continuously monitor building energy consumption, alert operations personnel early upon the onset of significant increase in consumption and assist them in identifying the problem.

In the six live building implementations, over eight building-years of operation, the "Cumulative Energy (or Cost) Difference" plot in ABCAT identified eight periods where significant energy consumption changes occurred that otherwise went undetected by the building energy management personnel. In the five retrospective building test cases, ABCAT detected 18 faults detected with "Days exceeding threshold" plots based on the simple standard that deviations greater than +/- one standard deviation (as determined from the statistics of the calibrated simulation) that persisted for a period of at least one month constituted a fault.

The potential future success of the ABCAT is strongly tied to the ability of future users to obtain accurate and reliable measurements. A strong emphasis in sound engineering practices of installation, data management, calibration and data prescreening must accompany the ABCAT to ensure verification of data quality, and the likelihood for success in implementing the tool.

In addition to the originally targeted goals of tracking and ensuring energy optimization in commissioned buildings, through the course of implementing and testing the ABCAT, several other added benefits or alternative functional approaches have been identified. These include use of the ABCAT as a commissioning savings tracking tool, a simple whole building energy analysis tool (even without the simulated consumption), and providing verification of, or use in filling missing metered or billing data, both important for customers of district utility providers, and the providers themselves.

4.2 The Diagnostic Agent for Building Operation (DABO™): a BEMS Assisted On-going Commissioning Tool[15]

4.2.1 Introduction

It is accepted that the initial, retro, or ongoing commissioning of HVAC systems are proven processes that reduces energy consumption and improves occupant comfort in buildings. Claridge et al. (1998) have shown that the use of the existing building control system for commissioning resulted in 25 % energy cost savings. In spite of documented benefits, commissioning is still regarded by many building owners as a minor activity in building operation and remains a one-time task that is performed during the building construction phase. Building professionals and owners attribute this situation principally to the cost associated with the commissioning process as well as the related difficulty of finding qualified resources to execute it.

The evolving capabilities of Building Energy Management Systems[16] (BEMS) can help to circumvent the barriers to commissioning by offering opportunities to automate some parts of the commissioning process. Automation or semi-automation of certain aspects of the commissioning process has the potential to reduce costs for commissioning, thereby leading to more widespread application of the process. Furthermore, automating this essentially manual process could allow its application on a regular basis, generating benefits over the entire life of a building. Developing a detailed systematic automated

---

[15] This section is abridged and edited from Choiniere and Corsi (2003) and Choiniere (2004).

[16] Building Energy Management System may also be referred to as an Energy Management Control System (EMCS)

approach will improve the quality assurance process and could even integrate energy audit capabilities that improve the overall performance of buildings. In this context, an on-going BEMS assisted commissioning tool, Diagnostic Agent for Building Operators (DABO) that verifies and optimizes the performance of building HVAC systems using the capabilities of BEMS was developed by the CANMET Energy Technology Centre. This tool is applicable mainly to commercial and institutional buildings. Commercialization is being considered.

On-going commissioning (IEA 2001a) is defined as a systematic approach used to inspect, verify and document the installations and operation of building systems to ensure that they operate at their optimum energy performance levels. This state is only achieved when buildings consume the minimum energy at the lowest cost while simultaneously considering the building's function and comfort level, available energy source(s), building energy systems and energy rates. To be efficient, many tasks must be performed continuously, an undertaking that can be facilitated by monitoring the condition of HVAC systems and building energy consumption using a BEMS.

Since 1998, the Canada Centre For Mineral And Energy Technology(CANMET) Energy Building has been used to test and demonstrate various tools developed in the context of the CANMET Intelligent Building Operating Technologies R&D plan (Jean, G. 2004) of which DABO is the central component. In this context and principally for the last eight years an on-going commissioning process has been conducted. DABO, a software package that uses a hybrid technology composed of conventional and artificial intelligence techniques to ensure optimum operation of building systems has been used actively in the project delivery system for the continuous monitoring of all HVAC equipment and meters (e.g., terminal unit, air handling unit, plant equipment and energy meters), the analysis of the incoming information, the detection and diagnosis of major HVAC component faults, non optimum set points and sequences of operation and the monitoring of implemented measures.

The Benefits obtained through the use of the DABO™ software are summarized as follows:

Energy savings	Improved performance of mechanical systems
Reductions in greenhouse gas emissions	Rapid and automatic fault detection
Improved occupant comfort	Energy performance monitoring
Capabilities for troubleshooting of malfunctions	Reduce operation costs

This chapter builds on the paper "A BEMS-Assisted Commissioning tool to improve the energy performance of HVAC systems" where Choinière and Corsi (2003) described the use of BEMS assisted commissioning tools and have identified the potential to facilitate the application of initial, retro and on-going commissioning processes.

In a subsequent paper ('Four years of On-going commissioning in CTEC-Varennes Building with a BEMS Assisted Cx Tool') Choinière (2004) presented results from the first four years of the on-going commissioning project performed in the CANMET ENERGY Building that generated a 35 % reduction in the energy used. DABO has largely contributed in the verification and optimization of the performance of the building.

This section presents the functions of DABO and the results of the first eight-years of on-going commissioning conducted at the CANMET Energy- Varennes building with DABO.

DABO, A BEMS-Assisted Commissioning Tool

DABO mainly includes two parts: the commissioning-assistant module and the fault management report.

The commissioning-assistant module, designed to assist and perform some functions described in the on-going commissioning process section is a module of DABO (Choinière 2001) which serves as the interface between the end-user (e.g., building operator, commissioning agent, and energy manager) and the control system (BEMS). As shown in Fig. 4.15, the tool continuously monitors the building control data and stores it in a structured database to be used on-line or upon request. Data resulting from standardized test procedures invoked manually or automatically are also stored in the database. The database functions as a server for reasoning algorithms that perform intelligent analyses of the monitored data, perform additional automated tests of components and systems, identify faults and diagnose them, and evaluate potential improvements in energy efficiency. The tool produces reports adapted to the different partners involved in the on-going commissioning process (building operators, service technicians, energy managers, commissioning agents, HVAC&R engineers).

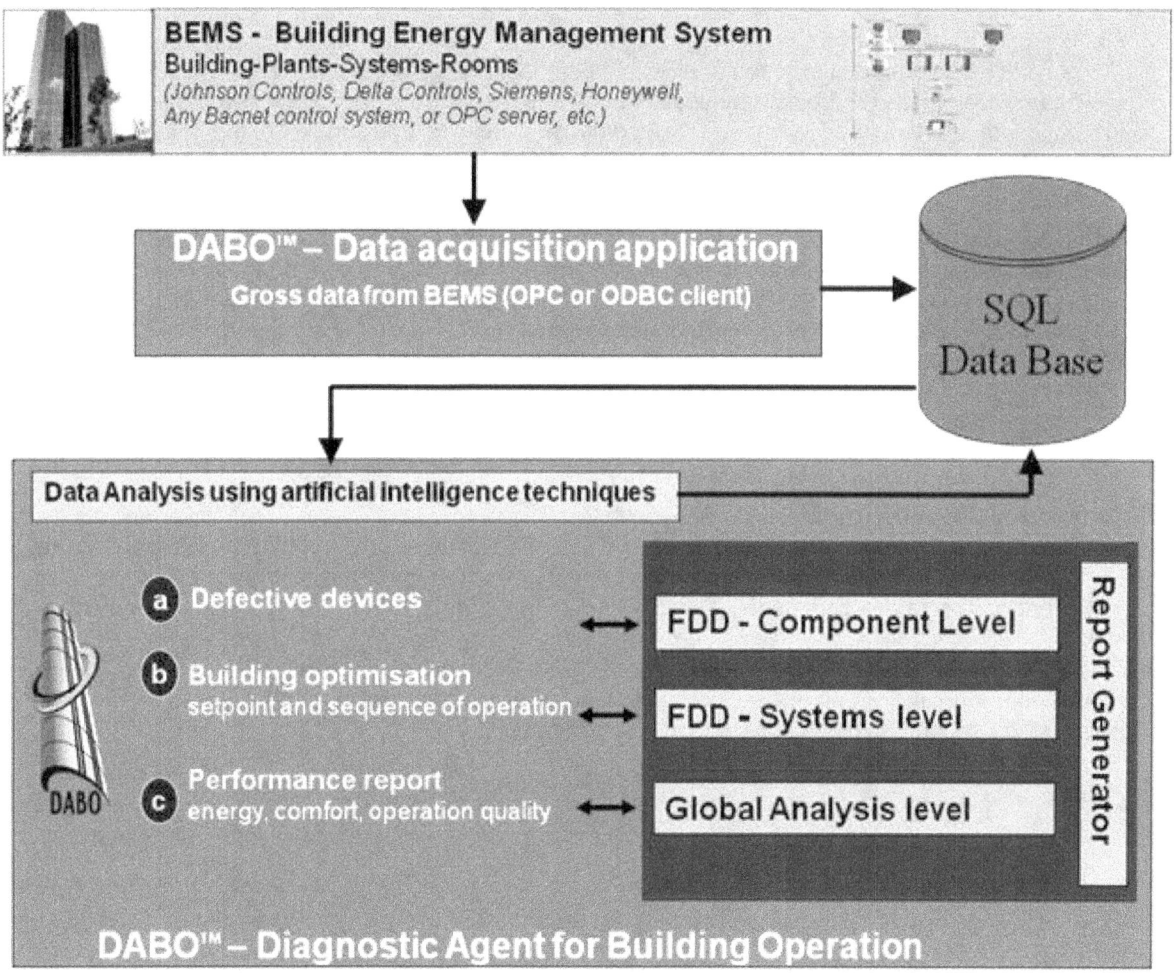

Figure 4.15 Structure of the On-going Commissioning Tool for HVAC Systems Embedded in DABO.

The standardized test procedures are performed at three levels.

- At the first level, an hourly component analysis of individual HVAC devices and equipment is performed automatically using a combination of control loop indices and expert rules to verify their proper operation.

- The second level of testing consists of an integrated system analysis to verify the operation and energy performance of the overall HVAC system over a longer period of time (e.g., hours, days, weeks or months). At this level a set of component performance indices and expert rules is also used in the analysis.

- The third level performs basic energy performance and operation control quality reports that provide the information required to evaluate potential energy measures on specific devices. To reduce data traffic on the communication networks, the tool's steady state detectors and zone fault detectors are directly embedded in controllers. Specific applications of the fault detection and diagnostic (FDD) methods implemented in DABO are described further in Section C of IEA 2001b.

The Fault Management Report (FMR) is a tool embedded into the DABO platform. The FMR report is used by the building operator/manager to manage information concerning confirmed faults in the building's operation detected by the DABO FDD module and the subsequence operational corrective actions needed or done. In this report the user can review actual and historical data for a specific device, for all devices of the same type or for all devices of a building.

Function of the information required, the FMR report can be displayed and printed in a summary or detailed form.

The following information is reviewed in FMR reports:
- Device type, System name, Fault number, Fault description, Fault start time, Confirmation date, Fault priority, Primary benefit, Secondary benefit, Impact on energy savings, Cost, Responsible operator, Operator's comments, Confirmation of repaired fault and Repair technician's comments

Figure 4.16 shows for all air handling units in a building a summary display of all faults detected that need to be repaired. Figure 4.17 is an example of a detailed 'repaired faults' report. This report is useful to produce a status report of actions done during a specific period.

System	Failure #	Failure description	Failure start	Confirmed date	Priority	Primary benefit	Secondary benefit	Energy savings	Cost ($)	Operator	C
M3 Bureaux 1	31	Point in manual	2007-04-20 08:00	2007-05-07 13:00	Extreme	Indoor environment	Operation and mai...	Yes		Daniel Choinière	c
M3 Bureaux 1	19	Control software problem	2007-02-19 05:00	2007-03-20 15:00	Moderate	Indoor environment	Operation and mai...	No			s
M2 Laboratoires	19	Control software problem	2007-06-19 06:00	2007-07-24 16:00	Moderate	Indoor environment	Operation and mai...	No		Daniel Choinière	s
M4 Corridor	19	Control software problem	2007-02-19 11:00	2007-03-21 12:00	Moderate	Indoor environment	Operation and mai...	No			c
M5 Salle électrique	14	Outdoor air damper failure	2007-03-14 09:00	2007-03-21 12:00	Low	Indoor environment	Operation and mai...	Yes			o
M5 Salle électrique	13	Mixed air damper failure	2007-06-08 00:00	2007-07-09 10:00	Low	Indoor environment	Operation and mai...	Yes		Daniel Choinière	A
M4 Corridor	13	Mixed air damper failure	2007-02-13 08:00	2007-03-21 12:00	Low	Indoor environment	Operation and mai...	Yes			c
M2 Laboratoires	12	Exhaust air damper failure	2007-04-15 21:00	2007-05-16 19:00	Low	Operation and mai...	Indoor environment	No		Daniel Choinière	ve
M30 Bureaux 2	128	Return fan over capacity	2007-05-29 14:00	2007-06-26 16:00	Low	Operation and mai...	Asset value or tena...	Yes		Daniel Choinière	c
M2 Laboratoires	6	Return air humidity sensor	2007-02-13 07:00	2007-03-21 12:00	Very Low	Indoor environment	Asset value or tena...	Yes			c
M3 Bureaux 1	35	Supply air CO2 sensor	2007-05-03 15:00	2007-05-09 15:00	Very Low	Indoor environment	Liability reduction	No		Daniel Choinière	s
M20 Annexe 2	116	Supply fan motor or drive failu...	2007-07-30 09:00	2007-07-31 12:00	Very Low	Indoor environment	Operation and mai...	No		Daniel Choinière	ve

Figure 4.16 Typical Fault Management Report for an Air Handling Unit in DABO ™

Repaired Failures Report

from '2004-10-06' to '2009-10-06'

Building: CTEC

Generated on: Tuesday 06 October 2009

2007/09/10	Medium	Supply air flow measurement	C: Eric Le Bouthillier	Sensor failure, defective power supply
			R: Eric Le Bouthillier 2007/09/14	installation of a new 4-20mmA power supply
2008/07/21	Medium	Cooling valve failure	C: Daniel Choiniere	Check if there is a leak on the cooling valve. On july 21, the valve was at 0% and SAT was at 16°C.
			R: Leonard Serravalle 2008/09/04	Linkage disconnected with the valve. Repaired linkage

Figure 4.17 Typical Repaired Failures Report for an Air Handling Unit in DABO ™

On-going Commissioning Project

The demonstration building is the CANMET Energy Building located in Varennes, Québec, Canada. Built in 1992, the single floor 3600 m^2 building includes office spaces for 90 people as well as two laboratories, two industrial pilot plants, conference rooms and a cafeteria.

The building, designed to be energy efficient, incorporates low energy technologies such as a passive solar preheating device, ice bank storage, photovoltaic cells, as well as a central gas heating plant and a central electric chilled water plant. Each area of the building is served by a specific air system designed for its occupation. The HVAC systems are central controlled by a BEMS system. (Table 4.3)

Table 4.3 CTEC-Varennes HVAC Systems

HVAC systems	Capacity	Location
Heating		
Fire tube boilers (2)	470 kW each	Building
1 primary and 5 secondary hydronic circuits, 7 pumps, constant volume		
Cooling		
1 air cooled chiller	406 kW	Building
2 ice bank tanks	1145 kW-hour	Building
1 hydronic circuit, 2 pumps, constant volume		
Air Handling system		
M1 (CAV, HEA)	2, 735 l/s	Pilot plant1
M2 (VAV, 100 % fresh air, heating, cooling)	5,815 l/s	Laboratories
M3 (VAV, heating, cooling)	5,500 l/s	Office phase 1
M4 (CAV, heating)	1,265 l/s	Storage phase 1
M5 (CAV)	160 l/s	Mechanical room
M6 (CAV)	1,030 l/s	Boiler room
M30(VAV, HEA, CO)	1,660 l/s	Office phase 2
M31(CAV, HEA)	5,200 l/s	Pilot plant 2
M32(CAV)	2,000 l/s	Mechanical room 2

The on-going commissioning process started in 1999 and continued until 2009 aimed at resolving operating problems, improving comfort, optimizing energy use and recommending retrofits where necessary. Delivery of the on-going commissioning project system included a series of tasks performed in four steps: planning, investigation, implementation and hand off (Table 4.4). Tasks surveyed with DABO are shown in italics. As it is an ongoing commissioning process, the investigation and implementation have been gradually and continuously performed over the 2000 to 2006 period. Since 2006, DABO is still used on a regular basis to insure the persistence of savings and detect new deficiencies.

Results for the investigation and implementation period 2000 to 2006 were presented in 'Four Years of On-going Commissioniong in CTEC-Varennes Building with a BEMS Assisted Cx Tool' (Choinière 2004) and are summarized in the following section.

Results for the hand off and persistence period include new deficiencies that occurred during the normal operation as well as deficiencies that were not detected during the commissioning of the various optimization projects (2006 to 2009).

Table 4.4 The Ongoing Commissioning Project Delivery System Followed at CETC-V

PLANNING
• Choose the team • Define project objectives, scope and deliverables • Review building documentation and energy bills • Develop Commissioning plan • Initiate cooperation with the building operation team
INVESTIGATION (continuous over 6 years, 2000 to 2006)

• Assessment
   o Site, design and occupant needs assessment
 	
• *Installation of DABO*
• *Develop and carry out diagnostic tests and system monitoring*
• *Analyze monitoring results*
• Develop list of deficiencies and improvements
   o Include capital improvement opportunities
   o Include training recommendations
• Select the most cost effective opportunities

IMPLEMENTATION (continuous over 6 years, 2000 to 2006)
• Implement improvements identified in investigation phase • *Retest and re-monitor to confirm the results* • Adjust, if necessary, the improvements carried out during the investigation phase • Review the energy consumption reduction estimates • Building Operator training and occupant information
HAND OFF- PERSISTENCE (continuous since 2006)

• Prepare and present final report
   o As-Built Re-commissioning work
   o New sequence of operation manual
   o Testing and balancing (TAB) report (air, water)
   o Energy baseline
   o Check-up of energy bills (3 months)
   o Proposal for EE measures with longer payback
• *Implement an on-going commissioning process and an energy management plan*
   o *Ensure that the use of DABO is well understood by the operators so as to maintain the re-commissioning benefits*

Results for Investigation and Implementation Step (1999 to 2006)

- Implement a continuous energy management plan 1998 (in house staff)
- Reset operation schedules (AHU, hydronic circuits)(1999)
- Optimise controls and sequence of operation
    - Function of actual needs
    - Peak load management (chiller, humidification)(2003)
    - Avoid simultaneous heating and cooling
- Reset set points (AHU, hydronic circuits)(2000 to 2005)
    - Minimum fresh air
    - Supply pressure and temperature
    - Night set back
- Fixed minor deficiencies
    - Sensor calibration
    - Low heating capacity in some rooms
    - Replacement of leaking valves
- Investment in measures with short payback
    - Addition of DDC controls (chiller, boiler 2001)
    - Link AHU M2 to solar wall (March 2003)
    - VSD on 3 fans (March 2002)
- Energy efficiency project (2005 to 2006) $250,000 CDN
    - Conversion of pneumatic room controls to DDC
    - Off peak electric boiler (200kW)
    - AHU M2 100 % fresh air convert to a recirculation system
    - Connection on AHU M4 fresh air to solar wall
    - VSD on 6 pumps and fan motors
- Energy reduction
    - Figure 4.17 shows the impact on the energy consumption of the on-going commissioning project implemented at the CETC-V since 1998. During this period 1999-2006, measures implemented have resulted in a 35 % reduction in electricity and 45 % in natural gas consumption.

Energy reduction

Figure 4.18 shows the impact on the energy consumption of the on-going commissioning project implemented at the CETC-V since 1998.

During this period, measures implemented have resulted in a 44 % reduction in electricity and 78 % in natural gas consumption. For 2007to 2008, the cost savings represented $91,861 CDN, or 51 % of the building energy bills . The Cumulative savings since 1998 are $496,124 CDN while the global energy used drop from 2248 MJ/m^2 to 994 MJ/ m^2.

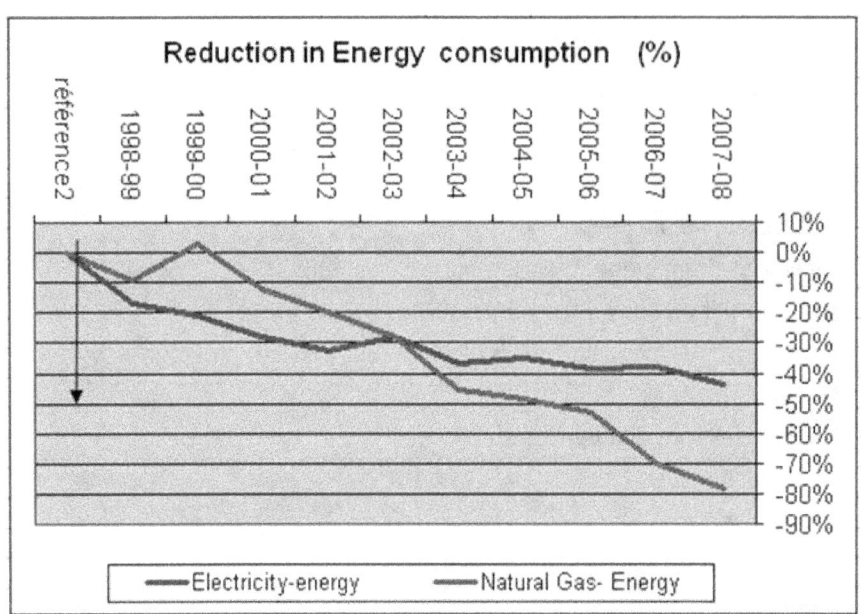

Figure 4.18 On-going commissioning impact on CETC-V Building energy consumption

How to Improve the BEMS-assisted Commissioning Tool (input from 11 demonstration sites)

The demonstration project was also used to validate the tool technology, to understand the barriers and opportunities and to identify areas for R&D activities to improve the prototype tool.

The following sections provide feedback collected from the tool installers, the building operators, Cx providers and facility managers on this project as well as on 11 other buildings where the technology has been deployed. This section also provides some thoughts on the lessons learned from our R&D projects.

Feedback from the installers (Total: 12 installations)

Perceived Strengths:
- Friendly user interface
- Most common HVAC system configurations can be set up
- Easy to install when the building information and connection to BEMS are available

Perceived Barriers:
- Installation is time consuming when building information is not adequate
- Remote accessibility (internet connection of the BEMS computer) is hard to get and frequently relatively slow. (Organizational Network department)
- Control procedures and control point naming conventions are not general (function of the control supplier and installer, the consultant, the owner)
- Lack and inaccuracy of the Documentation: (control and building drawings, balancing report)

Feedback from the building operators, Cx providers, facility managers (Total: 6 installations)

Perceived Strengths:
- Explanation and data supply for fault detected
- Powerful analysis of monitored data
- Very useful tool to secure persistence of savings

Perceived Barriers:
- The user needs basic computer knowledge
- The user needs a training period due to the number of available functions and the various acronyms used in tool reports
- Complexity to classify good and deficient devices (should have a summary report of defective components)

From facility manager, the information seems to be disparate.

Feedback from Tool developers

Perceived Barriers
- From the software aspect, DABO faced problems due to the obstacles inherent to the IT Office Network and the Building Energy Management Systems. The most common problems were: connectivity to Internet and/or control network and lack of resources from the owner for the installation of DABO.
- Also, the lack of availability and accessibility of the information on the mechanical systems and the sequences of control constrained the installation of DABO.
- From the human resources aspect, the barriers are related to the low level of knowledge of the mechanical system operator, and the insufficient time for training allocated to DABO and the turn-over of trained building operators.
- The absence of a consulting engineer and the activities related to re-commissioning pose a barrier to the use of DABO
- From a manager's point of view, the disparity and the quantity of information from approximately 20 different reports makes the day-to-day quantification of the tool benefits complex.

Opportunities
- The software prototype DABO is used and satisfies the user when the building owner accepts the tool and devotes satisfactory resources to ensure its installation and use
- Among the success factors we find:
  - The operator of the control system devotes his time to using DABO for the detection of faults, diagnosis and for EBCx measures. The time devoted to DABO varies according to the scope of the installation.
  - A consultant is hired to do the follow-up and effective management of project activities (The time devoted to DABO varies according to the scope of the installation)
  - The computer support staff is involved in the implementation from the very beginning to facilitate the set up of the DABO station and the connectivity to Internet for an external station.

- A consulting engineer is on-site and use DABO to carry out re-commissioning.
- The documentation on the mechanical systems and the sequences of control is available and up to date.
- The identified remedies for the detected faults are implemented.

Conclusions

An on-going commissioning process ensures that buildings achieve and operate at their optimized energy cost and performance levels, while ensuring comfort conditions for occupants. The project conducted at CANMET Energy has generated 44 % reduction in the electric use and 78 % in the natural gas consumption over the 1998 to 2008 period.

DABO, a BEMS-assisted commissioning tool has monitored the enormous amounts of data produced by BEMS and provided an extensive analysis of the incoming data.

The use of a BEMS assisted commissioning tool has helped to circumvent commissioning barriers by automating some parts of the process, which has reduced the costs for commissioning. Developing a detailed systematic automated approach has improved the quality assurance process and the overall performance of the building. Furthermore, automating this essentially manual process has allowed its application on an on-going basis, generating benefits over the entire life of the CETC-Varennes building.

The optimization process and the 'DABO^{TM}' tool are being demonstrated in more than ten projects. Demonstration projects include some of the first Canadian LEED buildings and the participation of major Canadian facility management firms and commissioning providers.

DABO^{TM} is a tool in constant evolution. Current research efforts are concerned with the development of an energy predictor, new fault detection and diagnosis modules for heating and cooling networks, and new analyses of BEMS data to enhance the commissioning process.

Demonstration projects are currently in the investigation and implementation stages; however early results show that DABO^{TM} helps circumvent commissioning barriers by automating some parts of the process, which has reduced the cost of commissioning. Developing a detailed systematic automated approach has improved the quality assurance process and the overall performance of the buildings. Furthermore, automating this essentially manual process has allowed its application on an on-going basis.

4.3 References

Energy Information Administration (EIA), 2007, Annual Energy Review 2007, Report DOE/EIA-0384(2007), U.S. Department of Energy, Washington, DC, June.

Choinière D. 2001. Un agent de détection et diagnostic de fautes pour les bâtiments. Congrès de l'Association québécoise pour la maîtrise de l'énergie. Quebec City, Quebec, Canada.

Choinière D, and M Corsi. 2003. A BEMS assisted commissioning tool to improve the energy performance of HVAC systems. Proceedings of ICEBO 2003. Berkely, CA.

Choinière D. 2004. Four years of On-going commissioning in CTEC-Varennes Building with a BEMS Assisted Cx Tool. Proceedings of ICEBO 2004. Paris, France.

Claridge D. E., M. Liu, W. D. Turner, Y. Zhu, M. Abbas, and J. S. Haberl. 1998. Energy and Comfort Benefits of Continuous Commissioning in Buildings. Proceedings of the International Conference Improving Electricity Efficiency in Commercial Buildings. September 21-23.

Claridge, D. E., W. D. Turner, M. Liu, S. Deng, G. Wei, C. Culp, H. Chen and S. Y. Cho. 2004. Is Commissioning Once Enough? Energy Engineering: Journal of the Association of Energy Engineering, 101(4): 7-19.

Curtin, Jonathan M., "Development and Testing of an Automated Building Commissioning Analysis Tool (ABCAT)" M.S. Thesis, Department of Mechanical Engineering, Texas A&M University, August, 2007.

Haberl, J. S. and E. J. Vajda. 1988. Use of Metered Data Analysis to Improve Building Operation and Maintenance: Early Results from Two Federal Complexes. *Proceedings of the ACEEE Summer Study on Energy Efficiency in Building, Asilomar, California.*

IEA. 2001a. International Energy Agency. Annex 40: Commissioning of Buildings and HVAC Systems for Improved Energy Performance. http://www.commissioning-hvac.org.

IEA. 2001b. International Energy Agency. Annex 34: Computer-Aided Evaluation of HVAC System Performance. Final Report. Editors Arthur Dexter and Jouko Pakanen. International Energy Agency.

Jean Gilles. 2004. A climate change solution; "Intelligent Building Operating Technologies". http://cetcvarennes.nrcan.gc.ca

Knebel, D.E. 1983. Simplified Energy Analysis Using the Modified Bin Method. Atlanta: American Society of Heating, Refrigerating and Air-Conditioning Engineers, Inc.

Lee, S. U. and D. E. Claridge. 2003. Field Tests of Whole Building Simulation for HVAC System Fault Detection, draft report of Energy Systems Laboratory, Texas A&M University July 2003.

Lee, S.U., Painter, F.L. and Claridge, D.E., "Whole Building Commercial HVAC Systems Simulation for Use in Energy Consumption Fault Detection," ASHRAE Trans., Vol. 113, 2007, Pt. 2, pp. 52-61.

Lin, G. and Claridge, D.E., "Retrospective Testing of an Automated Building Commissioning Analysis Tool (ABCAT)," Proc. 3rd Int. Conf. on Energy Sustainability, ASME, July 19-23, 2009, San Francisco, CA, USA, CD.

Turner, W. D., D. E. Claridge, S. Deng, S. Cho, M. Liu, T. Hagge, C. J. Darnell and H. J. Bruner. 2001. Persistence of Savings Obtained from Continuous Commissioning. Proceedings of the Ninth National Conference on Building Commissioning, Cherry Hill, NJ.

5. INTERNATIONAL COMMISSIONING COST-BENEFIT AND PERSISTENCE DATABASE

5.1 Introduction

It is generally recognized that demonstrating cost-effectiveness will remove a major barrier to the wider market acceptance of building Commissioning (Cx). Between 2005 and 2009, the U.S. team for the IEA Annex 47 led efforts to increase the diversity of Cx project data available and to quantify Cx costs and benefits for international Cx projects through Task 4 of the DOE-funded IEA ECBCS Annex 47 on Cost-Effective Cx of Existing and Low Energy Buildings, "*Technical Development and Data Population of the Cx Database (IEA Subtask C)*".

- The project collected **financial metrics** and **technical details** from international commissioning projects.

- The data is intended for use to demonstrate the value of commissioning to owners and possibly to assist in government policy-making.

Table 5.1 IEA Subtask C Objectives and Desired Impacts

Objective	Impact
Improve ability to estimate costs and benefits of future commissioning projects based on results of previous, similar projects	Increase adoption of commissioning by providing greater certainty about costs and benefits to building owners
Identify commonly occurring building problems or "hot spots" based on statistically significant data	Increase cost-effectiveness of commissioning by helping providers focus attention on likely problem areas Focus policy and technical R&D efforts on areas of opportunity for improving building energy efficiency and performance

Foundational work began in 2005 with an initial presentation of cost-benefit methodologies and output options at the Spring 2005 Annex meeting in Munich and a survey of Annex members on the subject of a data collection protocol at the Fall 2005 meeting in Prague (Friedman et al. 2005, see Appendix A Cost Benefit Protocols Report). At this meeting, Annex participants were engaged in collaborative planning for the project and set goals for expected participants' annual submissions of Cx and Existing Building Commissioning (EBCx) projects to an international Cx database, as shown in Table 5.2.

Work continued in 2006 with a study of Cx cost and benefit methodologies Friedman et al. 2005[17] and the first draft of the data collection tools (Excel forms). In 2006 and the first half of 2007 the data collection tools were continuously refined in collaborative processes with Annex members at the Spring 2006 Annex meeting in Trondheim, the Fall 2006 Annex meeting in Shenzhen, and the Spring 2007 meeting in Budapest, as well as the significant contributions of a specially formed Advisory Committee. The impact

---

[17] Friedman, H., M. Frank, T. Haasl, K. Heinemeier, 2006. Chapter 1 "State-of-the-Art Review for Commissioning Low Energy Buildings"

of the method used to determine savings persistence was evaluated and online testing of the automated persistence enhancement tool ABCAT (Automated Building Cx Analysis Tool) was carried out in four buildings. In Spring of 2007, the final data collection tools were released, previously received data was transferred from earlier versions of the tools and additional data collection efforts were engaged.

Over the following 18 months, opportunities and interest arose for additional supplementary funding to expand the project, first through an earmark to the National Center for Energy Management and Building Technologies (NCEMBT) which would have included the creation of an online searchable data base with many more U.S. projects. This expanded research opportunity diverted the progress of the initial project plan somewhat, as schedules and work plans were adjusted to accommodate the expected development of the online database including an online data submission tool over a 6 month period in the Summer and Fall of 2007. The scope of the project was revised again back to the original work plan, when it was discovered that the 2007 NCEMBT earmark would not be funded. Although the DOE has expressed strong interest as recently as July of 2008 in funding the proposed expansion of this project, including the creation of an online searchable Cx cost-benefit database for end users and significant U.S. project data additions from strategic market sectors, no additional work was funded during the Annex project.

5.1.1 Cost-Benefit Methodology

The cost/benefit methodology evolved through collaboration between the Annex participants between 2005 and 2007 (Refer to Appendix A for details). Through this process the goals and data collection format were developed, along with defining the types of data that would be collected relating to costs, energy-related benefits, and non-energy benefits. This process is defined in the project's Cost-Benefit Protocols Report (see Appendix A).

The output of this process was the creation of two Commissioning Data Collection Forms - one for new buildings and one for existing buildings. The surveys were divided into four sections:

- Project
    - Project Information
    - Technical Information
    - Cost Data
- Energy
    - Resource Savings
    - Resource Use after Cx (Persistence of savings data)
- Non-Energy Benefits
- Issues & Measures

There were "required" and "optional" elements. The required elements were intended to give enough data to perform a cost-benefit analysis, and therefore include fields such as "total cost of commissioning" and "annual energy savings from implemented measures".

Additionally, there were building characteristics that are required for analysis purposes such as floor area and year the project was completed.

5.1.2 This Chapter

This chapter, documenting the submission of data by participants in the IEA's Annex 47 and including an analysis of this project data, is the culmination of data collection performed from May of 2007 through November of 2008. These data collection efforts included email and phone requests for data, support of Annex members in supplying data using the data collection tools, follow up contacts with the data source to clarify questions related to data quality and for missing or incomplete data in required fields for cost-benefit analysis.

Overview of the information presented in this chapter:

- Summary of the most relevant findings in the body of this report, related not only to the Cx/EBCx projects knowledge gained, but also lessons learned about the process of collecting data for such a study.

- Discussion around what we've learned from this study and how it relates to existing knowledge about Cx industry and in the Cx cost-benefit area

- Recommendations for the path forward for international commissioning cost-benefit projects

- Appendices provide comprehensive analysis, data collection forms, and

- Interim reports created throughout the project

5.2 Data Collection

5.2.1 Data Collection Goals and Outcomes

At the Fall 2005 Annex meeting in Prague, participants agreed to goals for country submissions of new building Cx projects and existing building EBCx projects (See Table 5.2). At the Spring 2007 Annex meeting in Budapest, a schedule for submission of projects and overall goal presented called for 45 projects to be submitted by Annex members between December of 2007 and May of 2008.

Projects had been submitted slowly and infrequently by Annex members throughout this period, so a decision to extend the initial round of data collection was made by the U.S. team. A final round of data was received in the Fall of 2008, which brought the total number of projects submitted to 54. The final number of completed projects submitted by Annex members and included in this analysis are represented in Table 5.2. Of the surveys used to complete this study, many did not include data in all fields, and so the available dataset for analysis of each of the survey questions varies.

Table 5.2 Goals and Actual Submissions of Projects by Nov. 2008

Country	Goal Cx	Actual Cx	Goal EBCx	Actual EBCx	Total Projects
Belgium	N/A		N/A	6	6
Canada	2-6		2-6	2	2
Czech Republic	2-6		2-6		
Germany	2-6	1	6-12		1
Hungary	2-6		2-6		
Japan	2-6	6	2-6	7	13
Netherlands	N/A	1	N/A	9	10
Norway	2-6		0	5	5
USA	> 12	2	> 12	15	17
Total	24-48	10	26-48	44	54

5.2.2 Accuracy and Completeness of Data

Cx Data Set

Only ten new building Cx surveys were returned; Most surveys contained a completed Project Information section; however, very few of the returns included complete information on the remaining sections of the survey due to data collection limitations (see Section 5.4 Data Access Limitations for discussion).

In order to determine cost-effectiveness in terms of simple payback period of the submitted projects, the end results (savings) of the project are also required. There were six surveys with completed fields related to the estimated savings from Cx. Of these six, only four provided the complete cost data. Therefore, the cost-effectiveness of Cx was not able to be determined.   An underlying issue is the method for determining cost savings in new construction Cx. Three of the six surveys with completed Cx savings included the method used for determining the savings.

For the purposes of this analysis Cx costs were converted to US $ using the average exchange rate for the calendar year prior to the year of completion to estimate the timing when the decision to invest in commissioning was made.

EBCx Data Set

In examining the data from the EBCx surveys, it was found that while most projects had complete data for the project information section, fewer had the data required in order to determine cost-effectiveness (EBCx cost, pre- and post-EBCx and energy usage, energy unit cost). Simple payback was calculated for each project, which does not require conversion of currencies and allows comparison across the data set. Nineteen projects had sufficient data for calculating simple payback, and 12 of these were from the USA. As a result, this data cannot be considered a reliable basis for establishing international cost-effectiveness metrics. In this analysis, it is noted in those sections where data was not available from all projects and therefore a subset of the data was used.

For this analysis, EBCx costs have been converted to US $ using the average exchange rate for the calendar year prior to the year of project completion to estimate the timing when the decision to invest in commissioning was made.

5.3 Cost-Benefit Analysis Summary

5.3.1 New Construction Commissioning

The following section highlights the key findings following analysis of data collected on new construction commissioning:

Project Characterization

Ten new construction commissioning surveys from 4 countries were returned, as shown in Fig. 5.1. The combined floor area for all of the projects totaled nearly 2 million square feet (ft^2), with the smallest project at 9,000 ft^2 and the largest project at 508,853 ft^2. The wide variation in building size makes comparisons challenging, as the Cx process would be different for buildings at either end of the spectrum.

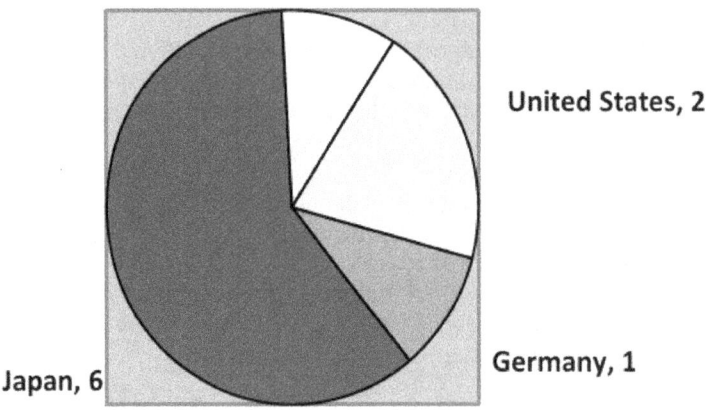

Figure 5.1 Breakdown of Cx projects by country

Owners were asked to rate their reasons for selecting Cx on a scale of 1 to 5 in a variety of categories –
with a score of 4 or 5, a factor is considered "important" . The results are summarized in Fig. 5.2.

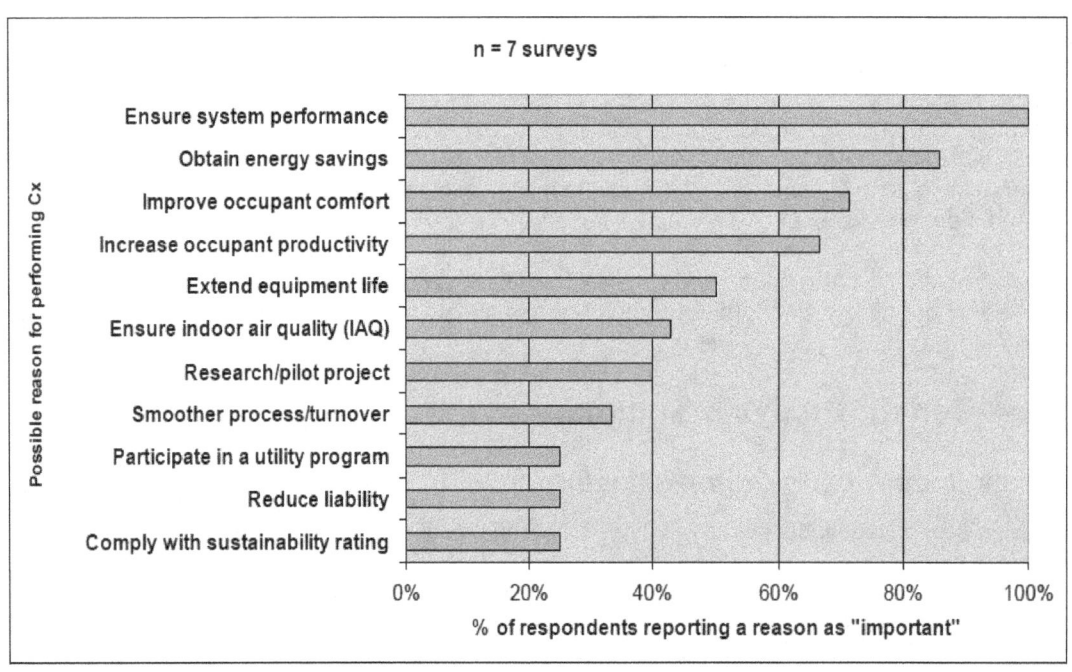

Figure 5.2 Owners' reasons for performing Cx

All respondents stated that ensuring system performance was important. The next highest factors were to
obtain energy savings, improve occupant comfort, and increase occupant productivity. It is interesting to
note that while ensuring system performance can be quantified, the other three factors are harder to
demonstrate for a Cx project.

Understanding the scope of the commissioning process undertaken provides a sense of how
comprehensive the commissioning process was for the seven projects that reported this information.
Figure 5.3 shows that scope varied widely across the small sample of projects.

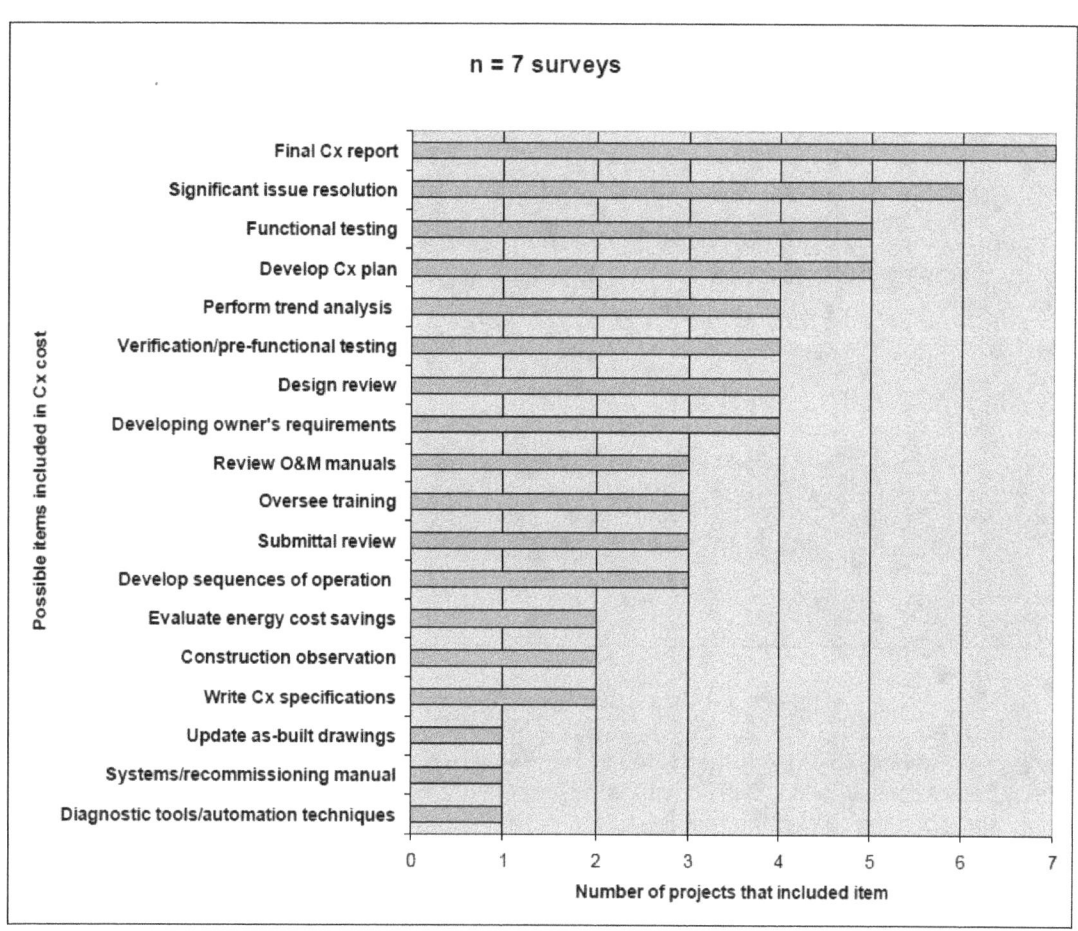

Figure 5.3 Items included in Cx cost

It is interesting to note that, while most building owners consider "obtain energy savings" as a main reason for performing Cx, only two out of the seven respondents included "evaluate energy cost savings" as part of their Cx process. This is likely due to the difficulties involved in evaluating energy savings for new construction projects, since there is no baseline energy use to compare against.

While no conclusions can be drawn with such limited data, the least often included items in scope (upated as-built drawing, systems/EBCx manual, and diagnostic tools) are consistent with our understanding of the U.S. experience. These items have high relevance for improving the persistence of benefits from commissioning.

Issues and Measures

A total of 55 issues compiled from six completed surveys were grouped by the type of system affected as shown in **Fig.** 5.4. Just over 65 % of the issues were found in the central plant and the air handling systems.

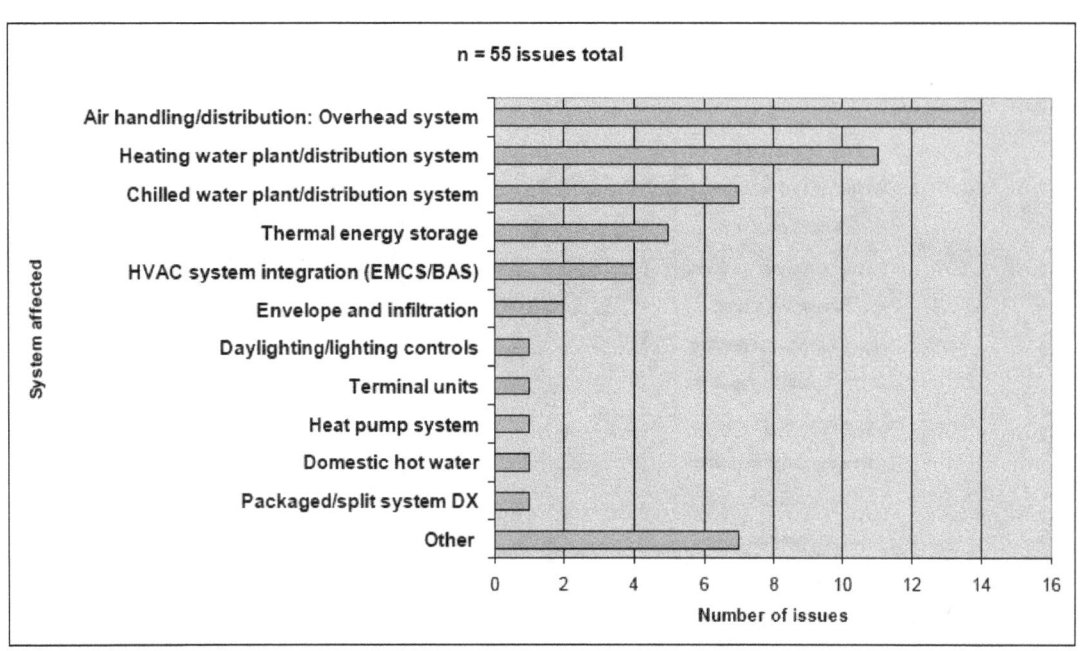

Figure 5.4 Cx Issues identified, by system type

Four surveys containing 45 of the issues listed in Fig 5.4 were classified according to four additional categories; design, construction, O&M or capital improvement. Figure 5.5 shows the breakdown of the additional categories (there were no responses for "capital improvement"). It is interesting to note the even split between 3 categories.

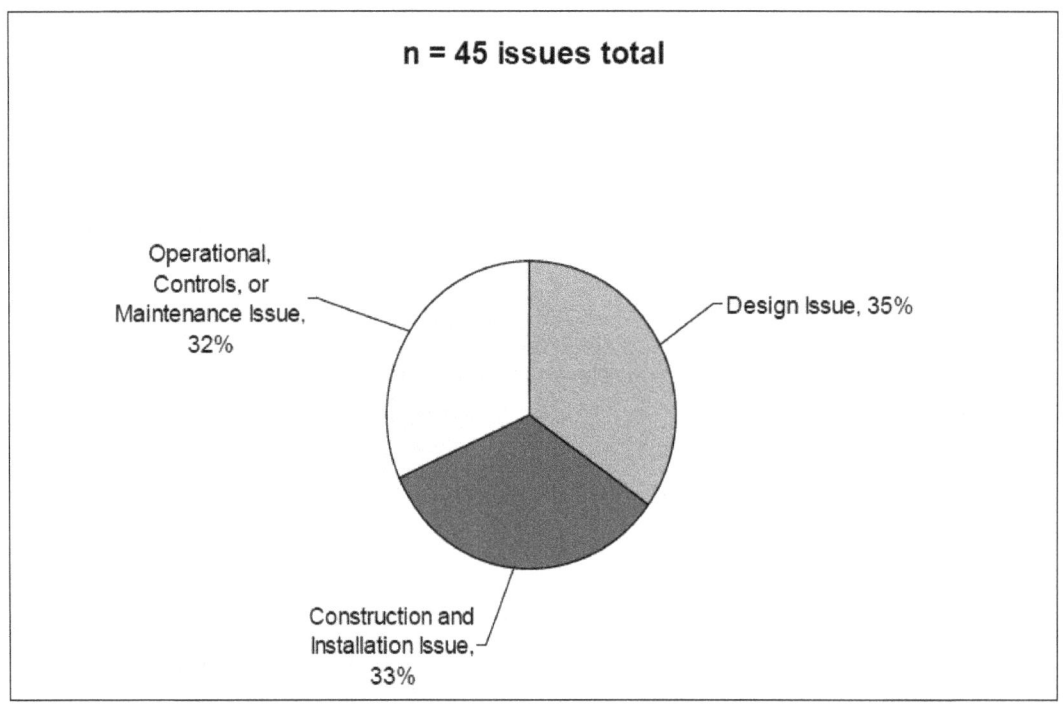

Figure 5.5 Cx Issues identified, by fault type

Out of ten returned surveys, only three had information associated with documenting issues and the measures implemented to fix them (see Fig. 5.6). In total, there were 56 reported measures, and all measures were reported as implemented except for a single electrical issue that was listed as unknown. The categories "Other," "Installation modifications," and "Design change" account for 84 % of all reported measures.

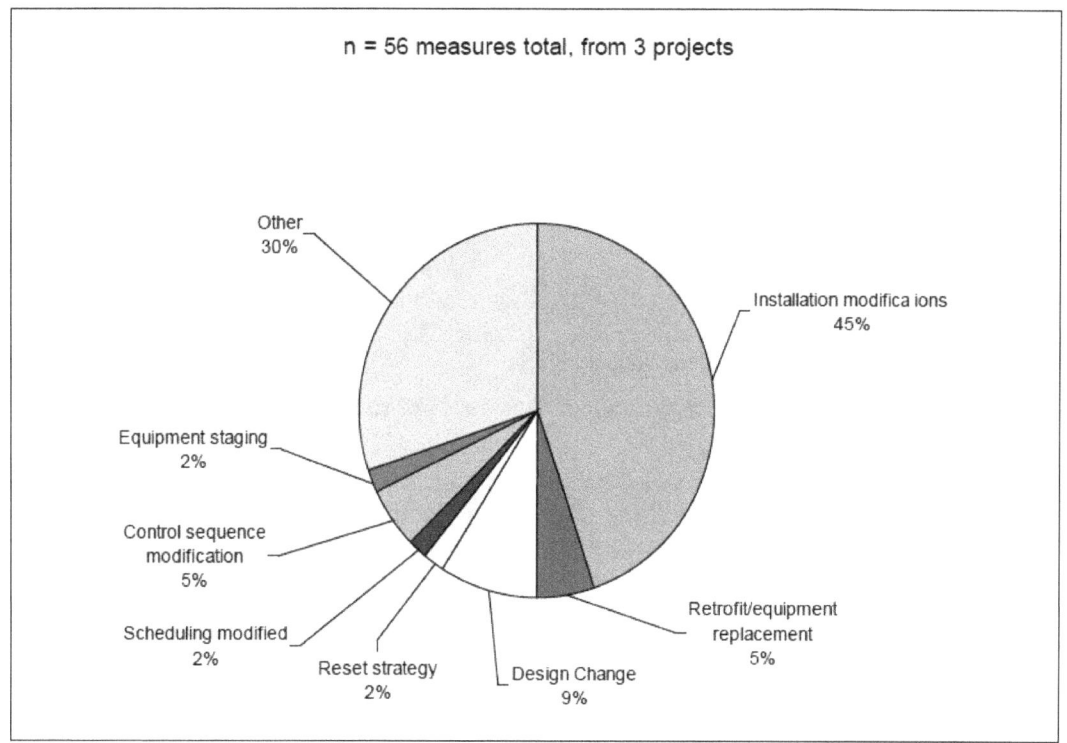

n = 56 measures total, from 3 projects

Other
30%

Installation modifica ions
45%

Equipment staging
2%

Control sequence
modification
5%

Scheduling modified
2%

Reset strategy
2%

Design Change
9%

Retrofit/equipment
replacement
5%

Figure 5.6 Implemented Cx measures

For the 17 survey answers with measure type "Other", the corresponding equipment categories were:

- Electrical (11)

- Security (2)

- 1 each for Ductwork, Fire/Life safety, Exhaust Fans, AHU distribution (overhead)

Costs

A total of six surveys were returned with costs reported in currency. Two additional surveys conveyed the total cost as hours of labor.

The Cx cost ranges from $0.06 per ft^2 to $2.57 per ft^2, suggesting the Cx process varied significantly and/or the way costs were attributed varies. This echoes earlier comments that the data set allows for little cross-comparison.

Three surveys were returned with a breakdown of fees. These results are shown in Fig. 5.7.

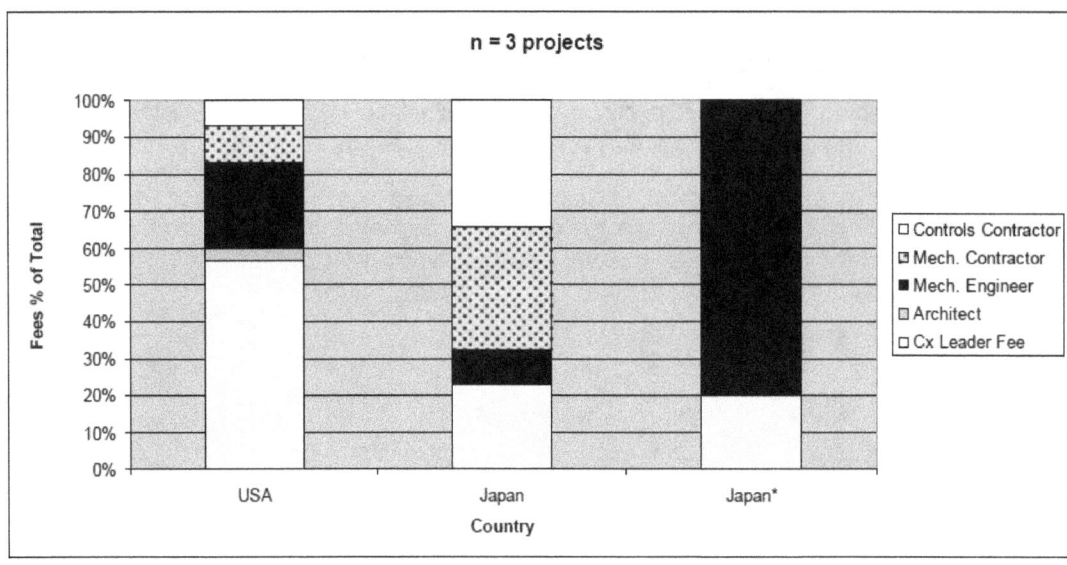

*costs reported as hours of labor for one of the Japan projects

Figure 5.7 Breakdown of fees as % of total Cx costs

The U.S. project was the only project to include architect's fees, and one of the Japanese projects only included fees for the Cx leader and mechanical engineering. For the U.S. project, the Cx leader fees account for over 55 % of the overall cost, suggesting a very active leadership role compared to the other projects where the Cx leader is perhaps facilitating/supporting the work of other parties.

Energy Savings

- Determining energy "savings" for a new building is challenging, and different methods may be used

- Claimed energy use may not account for the full floor area of a building, hence the EUI and savings percentage will not be accurately represented

5.3.2 Existing Building Commissioning (EBCx)

The complete results of the analysis of the existing building commissioning projects can be found in Fig. 5.8. The following section highlights the key findings.

Project Characterization

In total, six countries were represented in the data set. Significantly more U.S. buildings were available for inclusion in the data set, however it was decided that to retain a balance to the international data set, only 15 of these buildings would be entered.

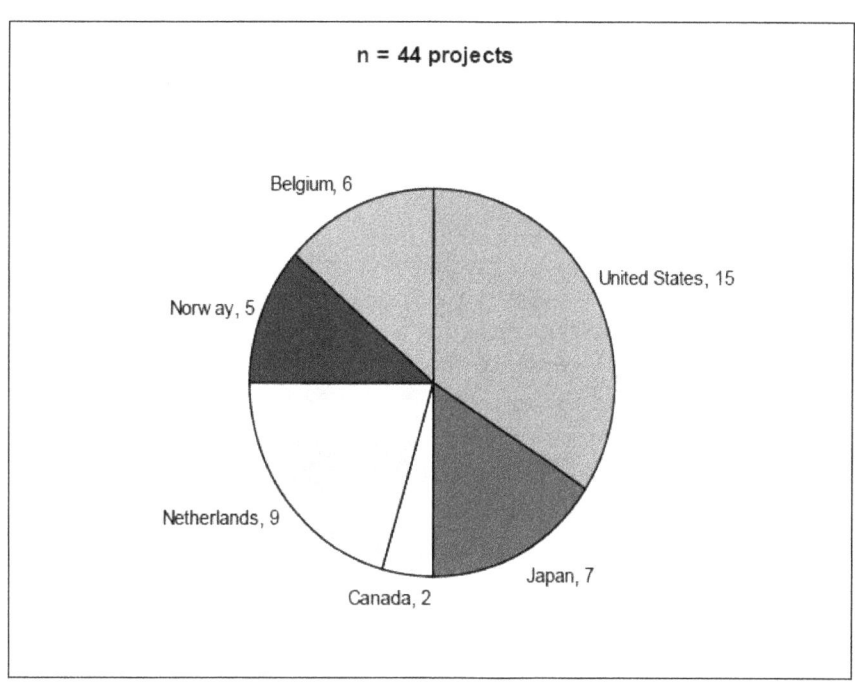

Figure 5.8 Breakdown of EBCx projects by country

The floor area from all of the projects totaled to more than 12 million ft^2, with the smallest building at 11,400 feet and the largest building at 1.4 million ft^2 The category with the greatest number of buildings is the range of 100,000 to 200,000 ft^2

The oldest building has a 1950 vintage, and the newest was built in 2006. Ten out of 44 buildings were less than five years old at the time of the EBCx project. This may be counter-intuitive for building owners, who feel that a relatively new building would not require system improvements, however it is consistent with the vintage of building often targeted in the U.S. for EBCx efforts since the control systems in these buildings often can accommodate greater energy-saving enhancements.

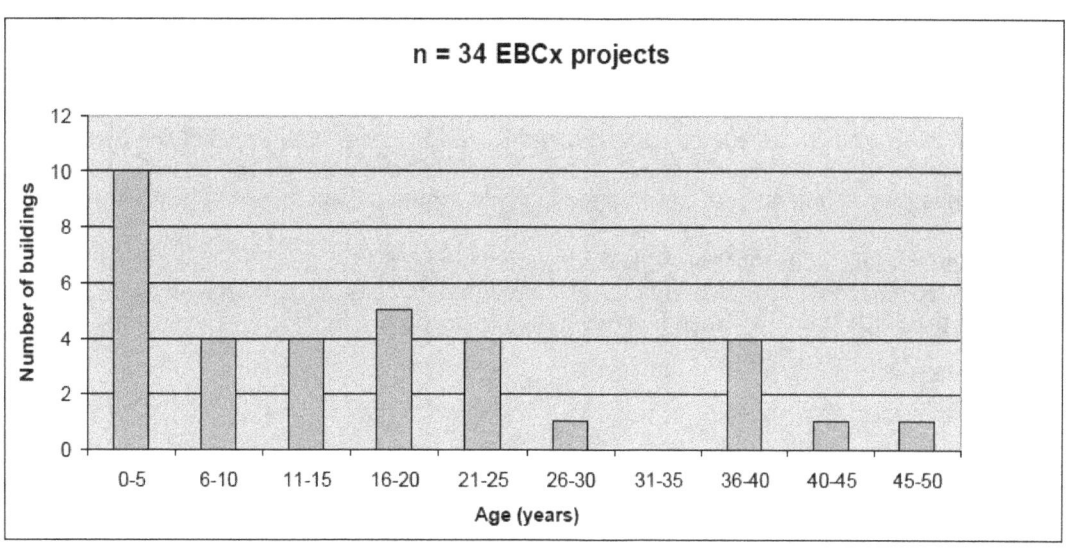

Figure 5.9 Age of building at time of EBCx project

Publicly owned buildings comprise 49 % of the data set; 51 % are privately owned. Of the publicly owned buildings, all but four are universities (one public assembly, one office, one laboratory, and one office). The majority of the buildings - 65 % - are owner-occupied, 28 % are leased, and 7 % are both owner-occupied and leased. Historically, public owner-occupied buildings are the early adopters of Cx efforts, so it is encouraging to note that private owner-occupied is gaining traction. Building leases often create disincentives for building energy improvements if all or a portion of energy costs are simply passed through to the tenants.

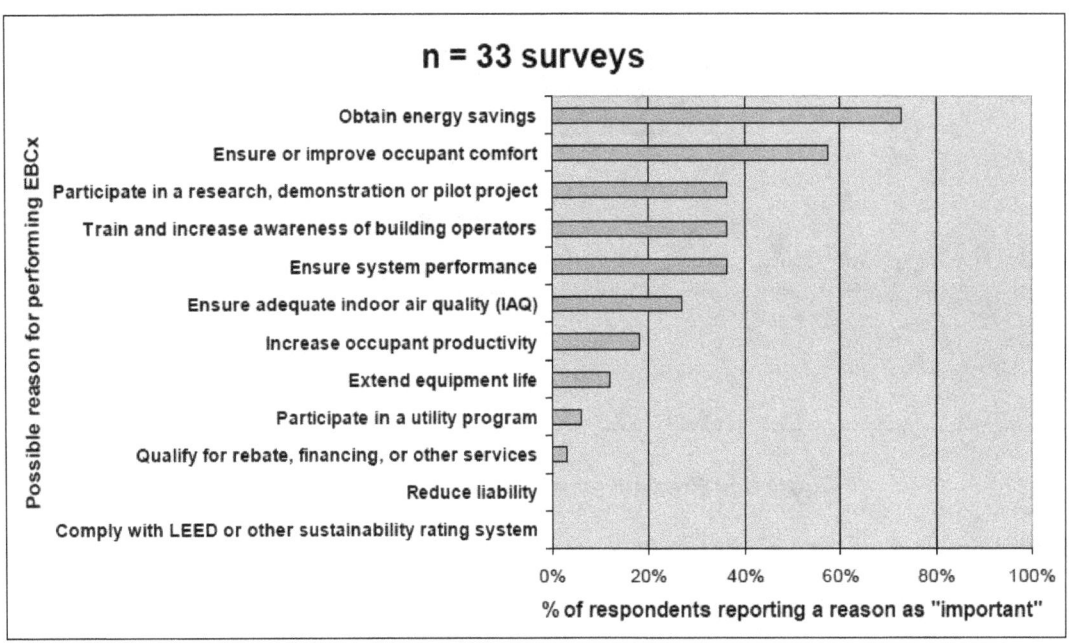

Figure 5.10 Reasons for performing EBCx

Internationally, the reasons for performing EBCx are consistent with expectations – to save energy, ensure occupant comfort, train operators, and ensure system performance. The fact that "participate in a research, demonstration, or pilot project" was also high-ranking reflects the state of the EBCx industry in many countries where EBCx is not an industry with practitioners, rather a research-grade effort.

Thirty-three out of 42 projects were led by an independent EBCx leader or energy service company. EBCx is still considered a specialized process, and so the skills have not yet been absorbed into conventional construction and mechanical engineering disciplines.

The full EBCx process includes most of the tasks included in Fig. 5.11, with a few tasks such as diagnostic tools/automation and implement capital improvements generally undertaken as an additional benefit beyond the EBCx process. Based on this data, the scope of EBCx generally includes:

- Trend analysis

- Document master list of findings

- Present a findings and recommendations report

- Implement operations and maintenance improvements

- Final report

Almost every project presented a findings and recommendations report, and of those that did not include this report, all but one included a final report instead.

The low occurrence of updating documentation, development of a systems/EBCx manual, and monitoring persistence is a result of the cost of these activities. The lack of documentation and monitoring may result in a lack of persistence of benefits from these EBCx projects.

One surprising finding about the scope of EBCx is that utility bill analysis, benchmarking, and calculating cost savings were only performed in about half of the projects. These are common EBCx activities in the U.S., but not as common in these international projects.

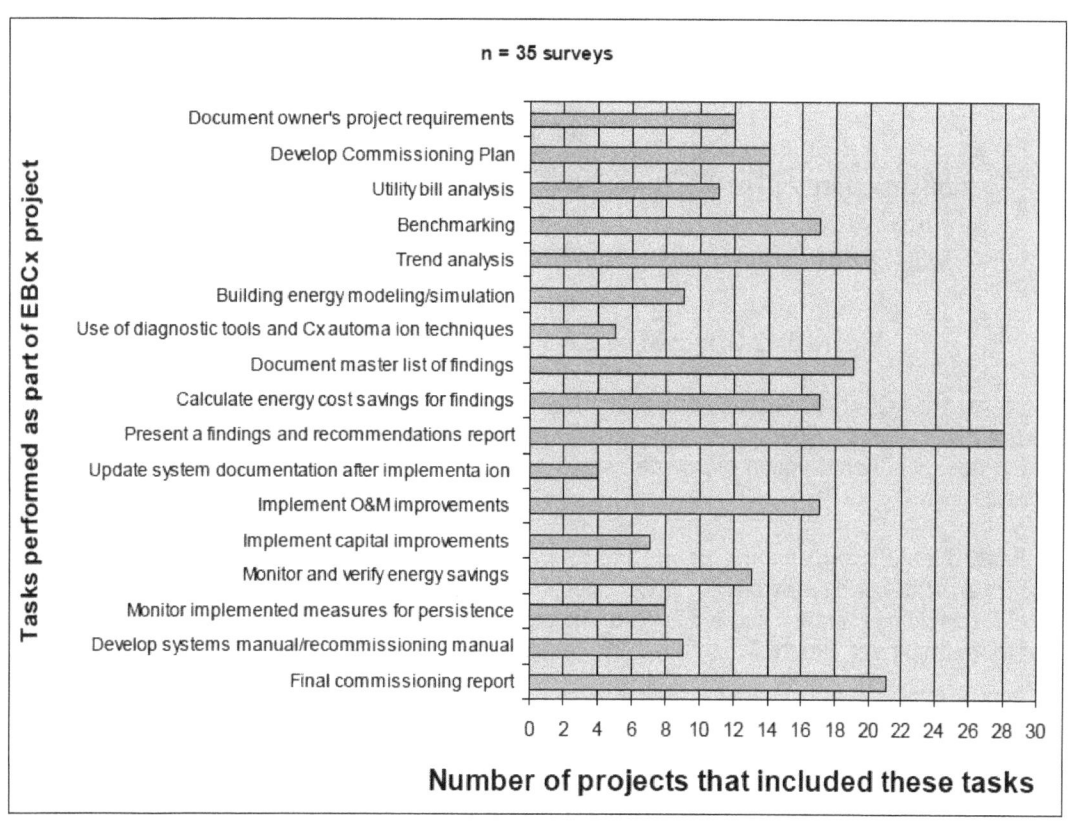

Figure 5.11 Tasks performed as part of EBCx project

Issues and Measures

The most common issues are as expected, related to integration and controls, air handling systems, and chilled and heating water plants. In order to get a better understanding of the impact of the phases of a building life-cycle on the problems found, a question was asked about where the issues originated.

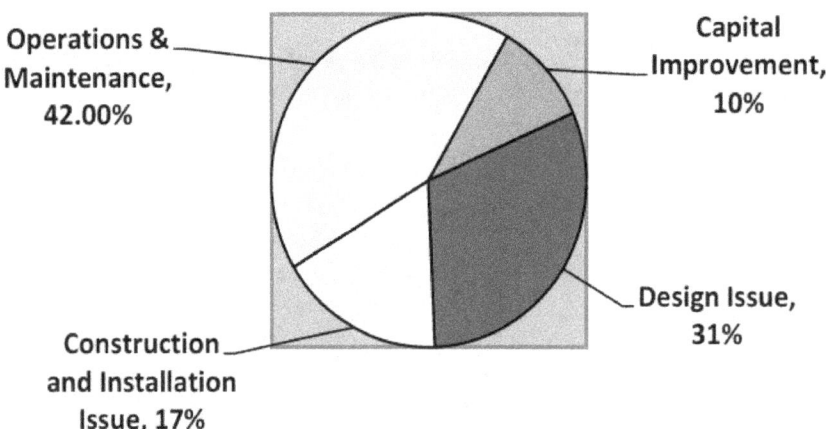

Figure 5.12 Origin of issues found through EBCx process

Here, the categories Operations and Control and Maintenance account for a large portion of the issues, areas where EBCx seeks to identify energy-saving improvements. Notably, 48 % of the issues were attributed to design or construction phase decisions, which points to the benefits of commissioning for new buildings.

In 38 projects, over 203 issues were reported through the survey and categorized by issue type (see Table 5.3 for a list of issue types available for selection by survey respondents). Design/installation issues included findings related to the design or installation of equipment, rather than the operation and control of that equipment. Figure 5.13 shows the results from completed surveys for 38 EBCx projects.

Table 5.3 List of issue types available for selection by survey respondents

Design/Install
Design detail inadequate
Equipment selection inappropriate
Lighting - spaces over-lit
O&M access insufficient
Overpumping or throttled discharge valve
System selection inappropriate
Equipment not based on design
Equipment not properly installed
Maintenance issues
Ductwork leaky
Filtration requires modification
Flow obstructions
Poor actuator operation
Valves leaky
Scheduling
Equipment scheduling sub-optimal
Equipment staging sub-optimal
Equipment start/stop sub-optimal
Lighting scheduling sub-optimal
Controls
Control loop needs tuning
Manual changes or overrides causing problems
Sensor problem
Sequence of operations inadequate
Simultaneous heating and cooling
Outside air
Economizer sub-optimal
Ventilation issues
Reset Strategy
Reset strategy: pressure reset strategy sub-optimal
Reset strategy: temperature reset strategy sub-optimal
Retrofit
Fan motor - no VFD
HVAC equipment inefficient
HVAC system inefficient
Lighting inefficient
Setpoints
Setpoints sub-optimal
VAV box minimums high
Variable flow
Variable flow: pump speed/flow high or constant when it should vary

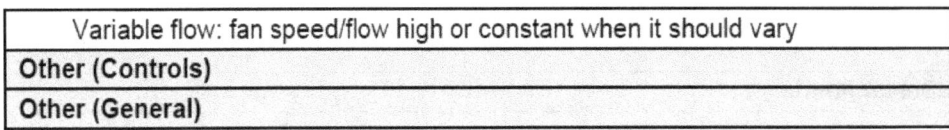

Variable flow: fan speed/flow high or constant when it should vary
Other (Controls)
Other (General)

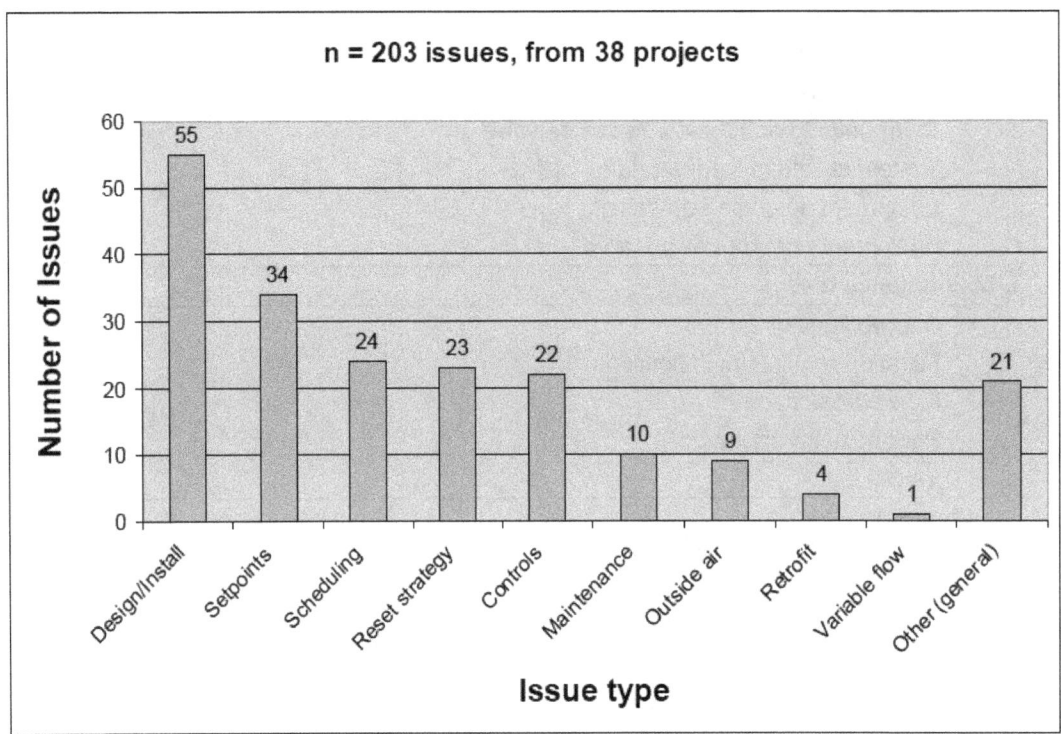

Figure 5.13 Issue types uncovered through EBCx process

One of the most unexpected findings in this report is that "Design/Install" was clearly the most common issue. This issue broke down into 9 sub-categories that were selected as follows.

- Construction/installation (24 issues)
- Equipment selection inadequate (15)
- Design detail inadequate (8)
- Space over-lit (2)
- Not properly installed (2)
- O&M access insufficient (1)
- Over-pumping/throttled valve (1)
- System selection inadequate (1)
- Equipment not based on design (1)

This finding points to the need for design and construction-phase commissioning of new buildings, as the problems can plague a building for its life, and finally be addressed in the EBCx process.

Figure 5.14 shows the systems affected for the 203 reported issues, with more than half of the issues being attributed to the "Air handling/distribution: overhead" or "Multiple HVAC systems" categories. These responses are generally in line with expectations.

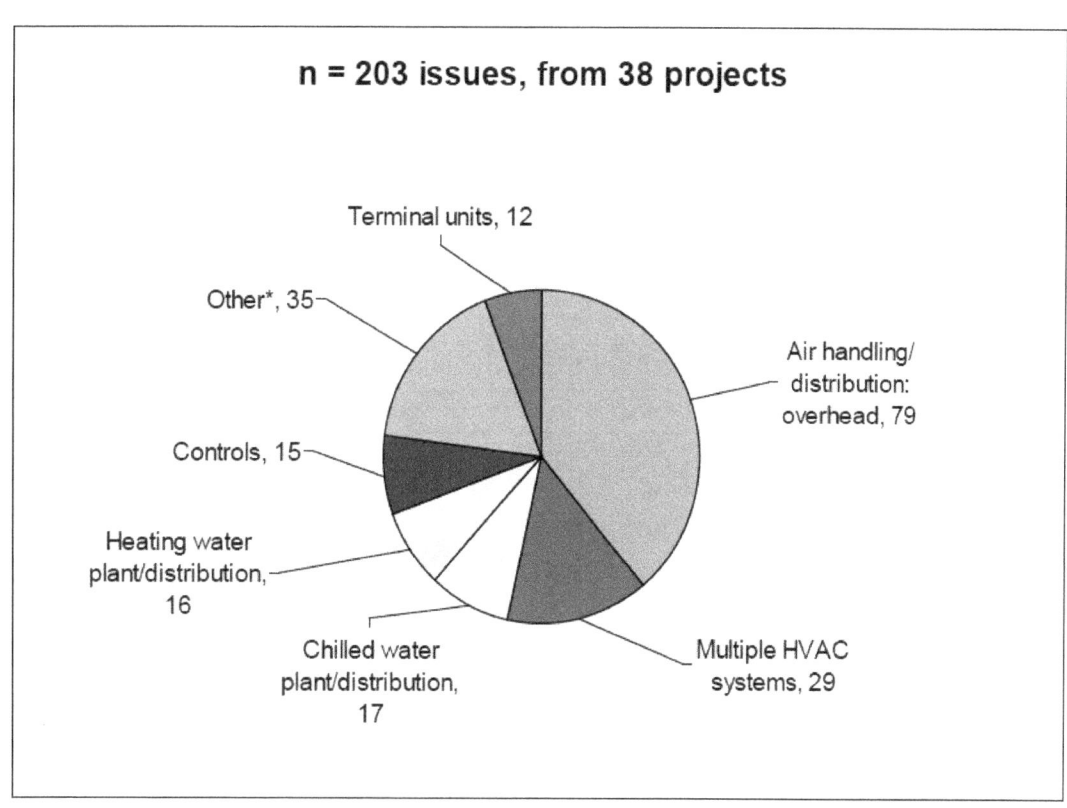

n = 203 issues, from 38 projects

Terminal units, 12

Other*, 35

Air handling/ distribution: overhead, 79

Controls, 15

Heating water plant/distribution, 16

Chilled water plant/distribution, 17

Multiple HVAC systems, 29

Figure 5.14 EBCx issues, system affected[18]

* Category "Other" includes lighting/daylighting controls, radiant heating, heat pump system, envelope/infiltration, plumbing, radiant cooling, thermal energy storage, all of which received 5 or fewer responses.

In addition to reporting the issues/problems identified through EBCx, survey respondents also reported the recommended fixes/measures for addressing those problems. Respondents gave the measure type, and for each measure indicated whether it was implemented, not implemented, or "unknown" (see Fig. 5.15).

Out of 205 recommended measures, 121 were implemented (59 %). While 59 % of the recommended measures were implemented, it is not known what percentage of available energy savings this represents. It is assumed that measures with higher savings would be more likely to be implemented and that the achieved savings would be greater than 59 % of the potential savings, but this cannot be verified for this study.

---

[18] Each category shown in this chart *includes* its associated controls. The standalone category "controls" includes general controls issues that did not fit into any other specific system type.

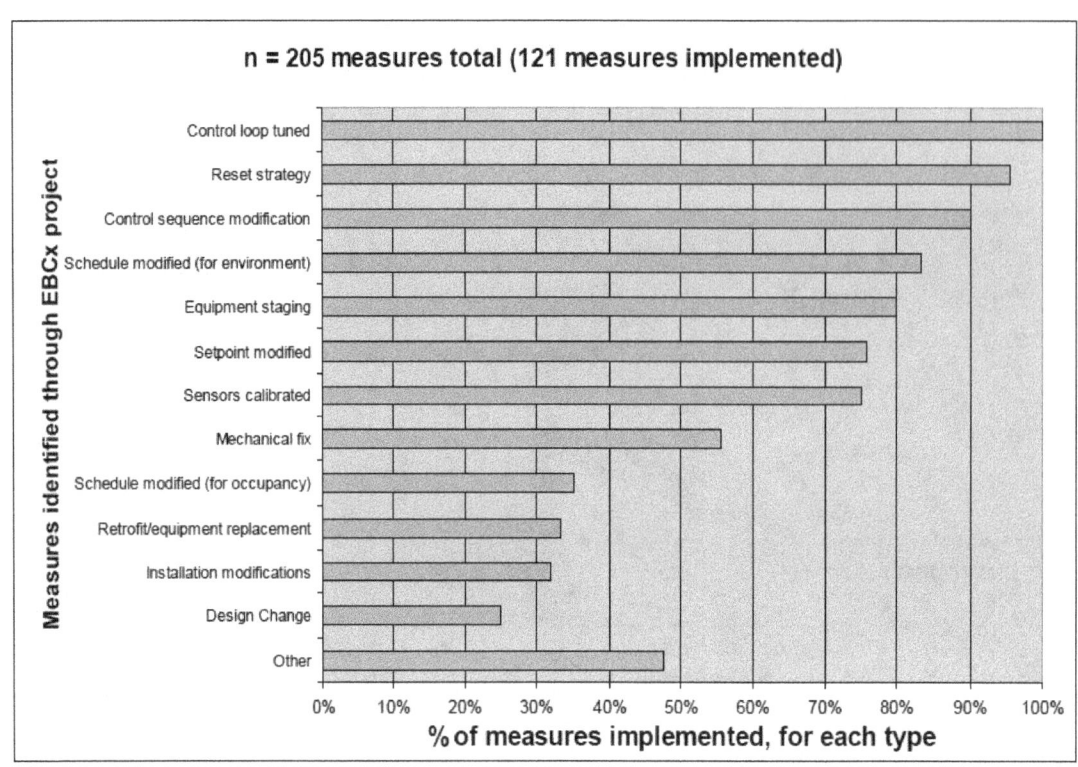

n = 205 measures total (121 measures implemented)

Measures identified through EBCx project

- Control loop tuned
- Reset strategy
- Control sequence modification
- Schedule modified (for environment)
- Equipment staging
- Setpoint modified
- Sensors calibrated
- Mechanical fix
- Schedule modified (for occupancy)
- Retrofit/equipment replacement
- Installation modifications
- Design Change
- Other

0% 10% 20% 30% 40% 50% 60% 70% 80% 90% 100%

% of measures implemented, for each type

Figure 5.15 Percentage of measures implemented, by type

This data shows that a relatively small percentage of the retrofits/equipment replacements and installation modifications were implemented, whereas the majority of the setpoint modifications, reset strategies, and other modifications to the control sequence of operations were implemented. This is in line with expectations, as the items that were implemented are generally lower cost than those that were not implemented.

Costs

The cost of commissioning existing buildings in all countries was normalized into US$/ft^2 and US$/m^2 using the average exchange rate for the year that the EBCx project was completed. To convert from US$/ft^2 to US$/m^2, multiply by a conversion factor of 0.0929. Through this analysis, median cost was found to be $3.12/m^2 (US$0.29/ft^2), with a range of US$0.11/m^2 to US$17.76/m^2 (US$0.01/ft^2 to US$1.65/ft^2) (see Fig. 5.16). It is important to note that cost data was not always available, especially for research-grade projects. In research projects, it is often difficult to separate out what might be deemed a typical commissioning scope.

Notes on projects with the highest and lowest unit costs per ft^2:

Highest unit cost:
Canada, EBCx cost US$/m^2 (US$1.65/ft^2)

- ° 465 m^2 (5,000 ft^2) 65 % office, 35 % "Research Areas"

- ° The list of tasks included in the process denote this as an example of a comprehensive EBCx project

- ° Performing comprehensive EBCx on a small building with research areas would understandably result in a high unit cost per ft^2

Japan, EBCx cost US$9.58/m^2 (US$0.89/ft^2)

- 9383 m^2 (101,000 ft^2), office building

- There is nothing in the building size, type, or EBCx tasks list to suggest why this building had a high unit cost per ft^2

Lowest unit cost:
Netherlands, EBCx cost US$0.11/m^2 (US$0.01/ft^2)

- 19974 m^2 (215,000 ft^2), university

- Project cost only 2,000 Euros

- EBCx scope included only trend analysis and a final report

- The limited scope explains the low cost

Netherlands, EBCx cost US$0.65/m^2 (US$0.06/ft^2)

- 30935 m^2 (333,000 ft^2), 80 % office, 20 % restaurant

- This project also had a limited scope

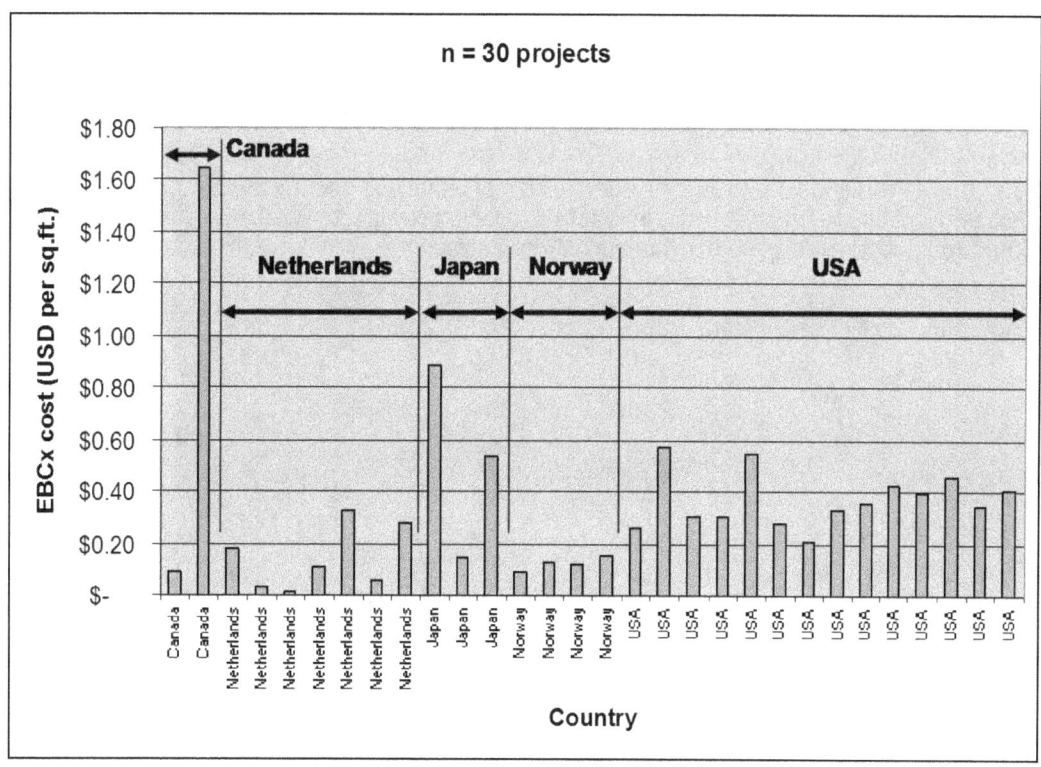

Figure 5.16 EBCx projects' cost per ft[219]

---

[19] To convert from US$/ft^2 to US$/m^2, multiply by a conversion factor of 0.0929.

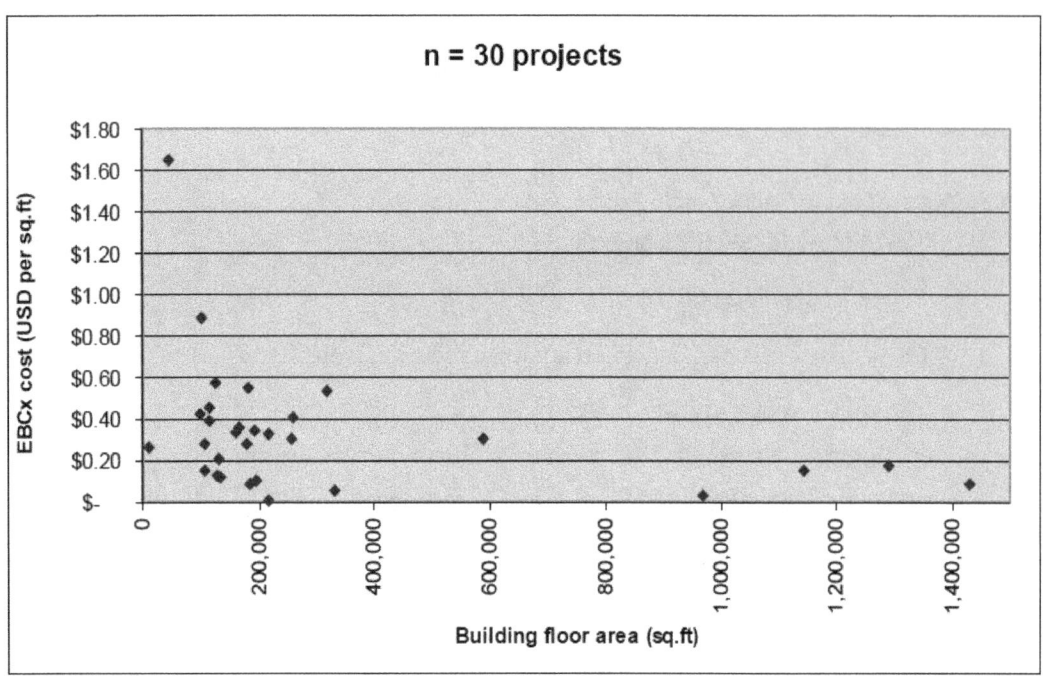

Figure 5.17 EBCx project cost per ft² vs. building size

Figure 5.17 shows a slight correlation between the size of a building and the EBCx cost per ft², but a larger dataset would be required in order to create a reliable correlation. Within the size range of 4645 m² to 37160 m² (50,000 ft² to 400,000 ft²), the scatter looks random.

---

[20] To convert from ft² to m², divide by a conversion factor of 0.0929.

Energy Savings

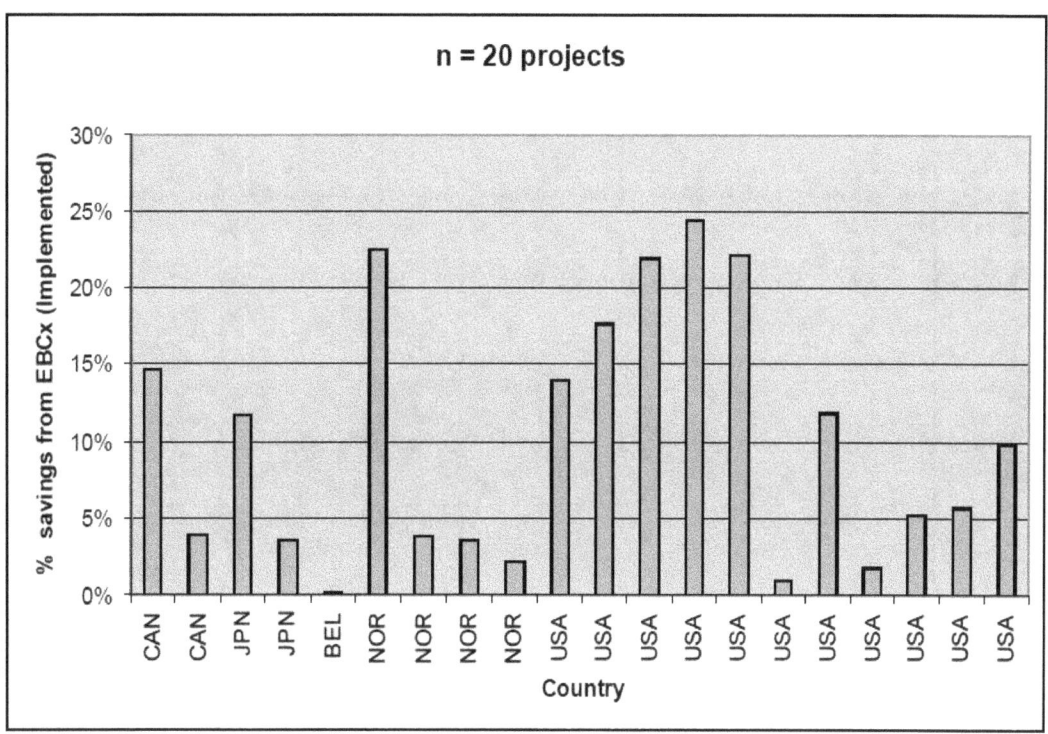

Figure 5.18 Percentage electric savings from EBCx projects

The data in Fig. 5.18 show a range of 1 % to 24 % electric savings (kWh), with a median value of 8 %. There has been no investigation into the sources of variation in these reported savings, but they could be significantly affected by the following factors:

- Scope of EBCx project

- Budget limitations resulting in efficiency improvements not being implemented

- Accuracy of estimates for claimed energy savings

- Energy efficiency of building systems prior to EBCx project

Energy Use Intensity (EUI) is one measure that *may* indicate the energy efficiency of building systems. Electric EUI is measured in kWh/ft^2/yr, and Fig. 5.19 plots electric EUI against implemented kWh savings for this data set.

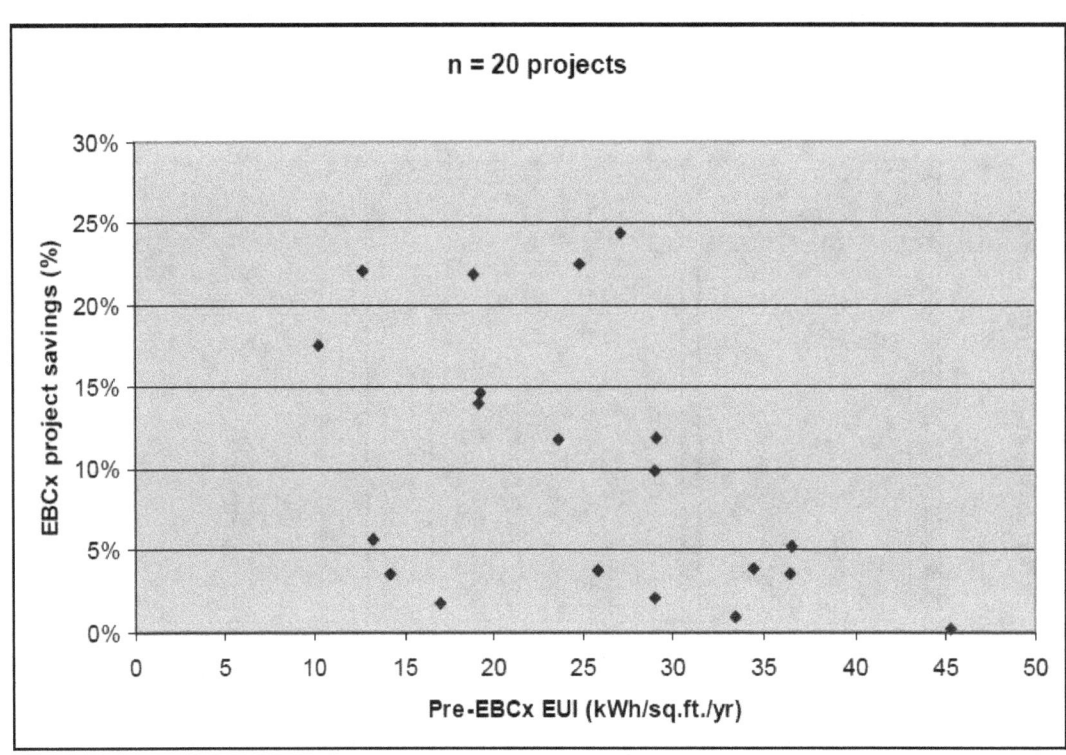

Figure 5.19 Electricity consumption vs. % savings for EBCx projects

It is logical to assume that buildings with a higher EUI would have greater savings potential, but Fig. 5.19 suggests that there is no correlation between pre-EBCx energy use and EBCx savings percentage for this data set.

Cost-effectiveness

Common measures for cost-effectiveness are simple payback and ROI (return on investment). For this study, simple payback is used as the cost-effectiveness metric, and is calculated as the overall project cost divided by the claimed annual cost savings. A 2004 report summarizing data from 100 projects in the USA showed a median simple payback of 0.7 years, with values ranging from less than 1 month up to values in excess of 5 years. A summary of 19 projects from this study are included in Fig. 5.20.

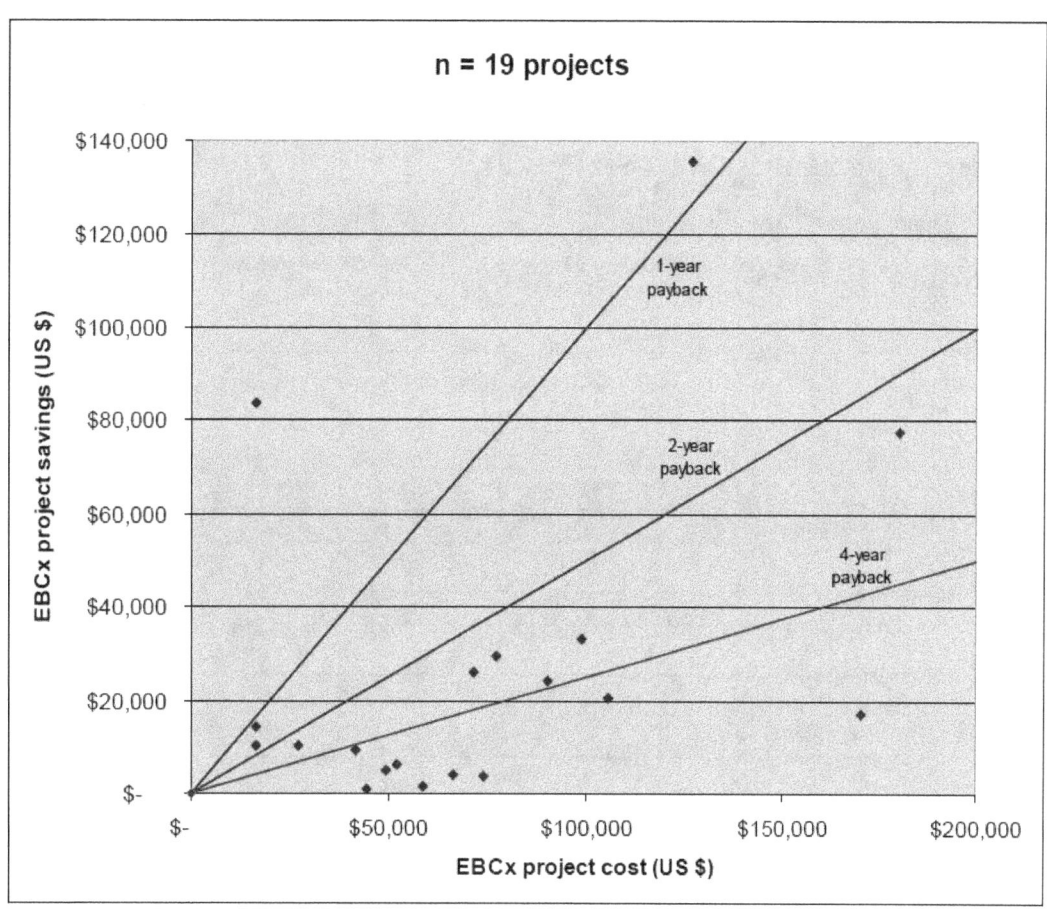

Figure 5.20 **EBCx project costs vs. annual savings estimates**

Project simple payback values ranged from 0.9 years to 45.7 years, with a median value of 3.7 years. Nine out of the 19 projects had a payback of greater than four years, and six had payback of between two and four years.

Higher payback is the result of *either* relatively high EBCx costs *or* relatively low resultant cost savings. For cases where payback was greater than four years in this study, the savings were relatively low (based on savings per ft^2), as opposed the costs being relatively high.

While this data is of interest and is useful as the basis for further study, it should not be used as the basis for predicting EBCx project savings, for a number of reasons:

- Data is from a relatively small number of projects

- There can be a wide variation in the way that cost savings are calculated, and they may not be comprehensive (e.g.. Electric savings calculated but not gas)

- Some projects may have been undertaken in order to address non-energy issues

- The scope of EBCx varied between projects

Non-Energy Benefits

Thirty-nine surveys were completed relating to non-energy benefits; these surveys were split under three headings, as shown in the following figures:

- Operations & Maintenance (O&M) – See Fig. 5.21.

- Indoor Environment – See Fig. Figure 5.22.

- Asset value and liability reduction – See Figure 5.23.

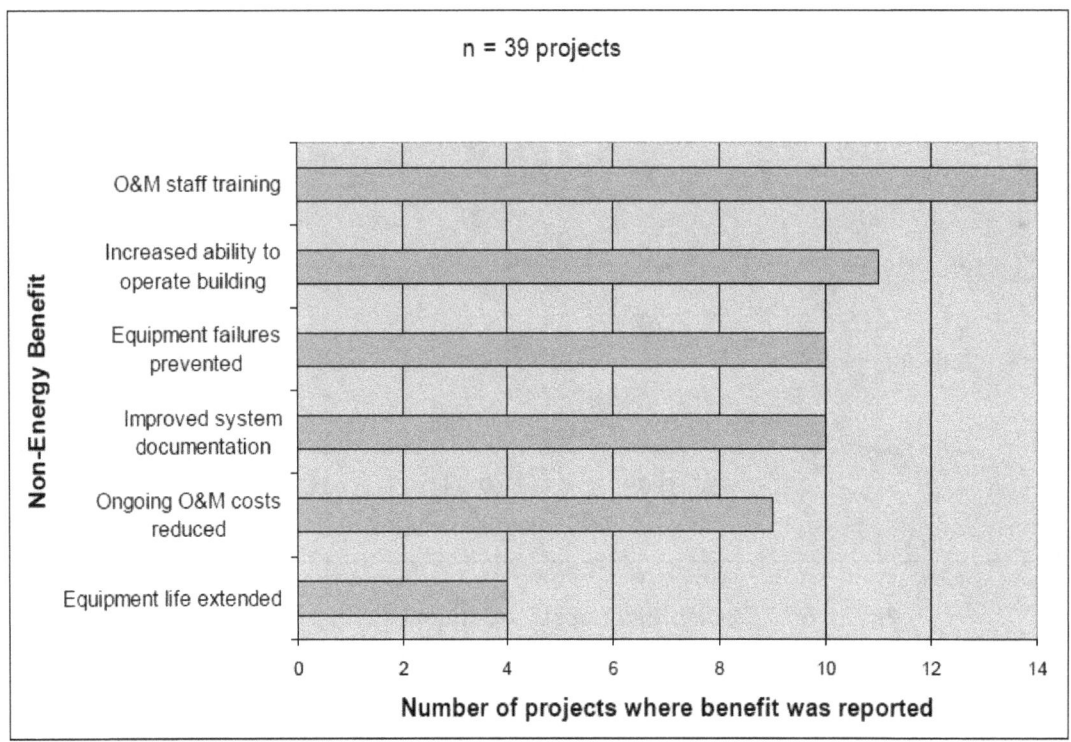

Figure 5.21 Non-energy benefits – operations & maintenance (O&M)

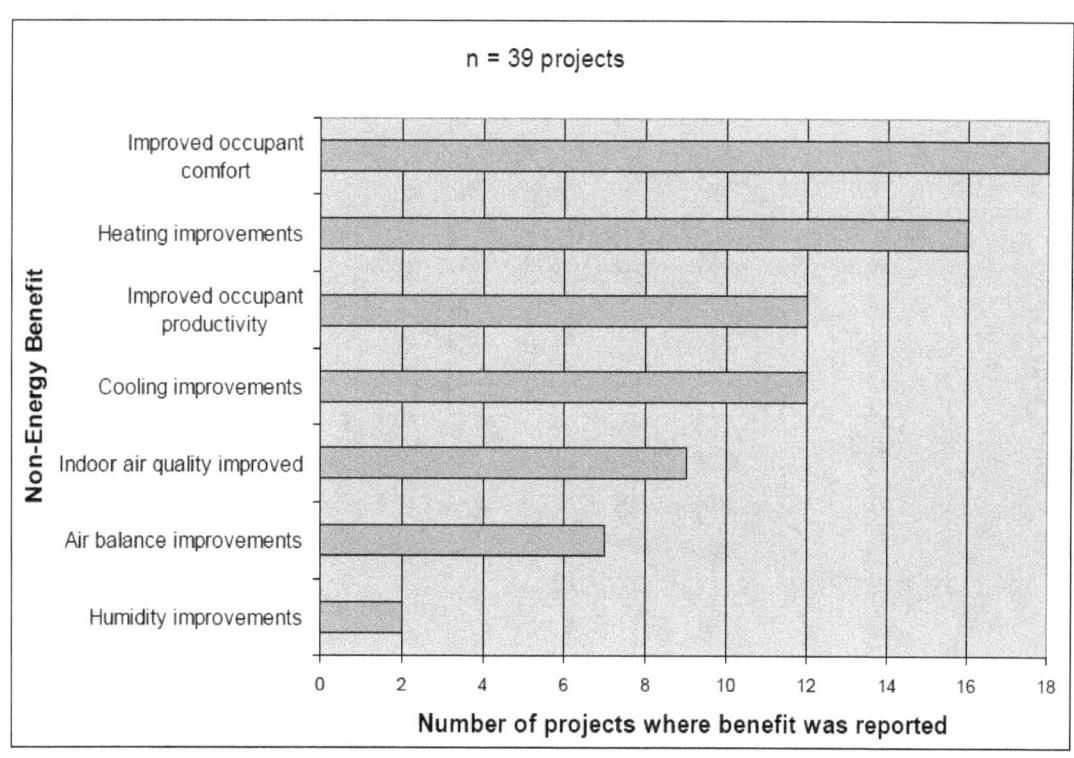

Figure 5.22 Non-energy benefits – indoor environment

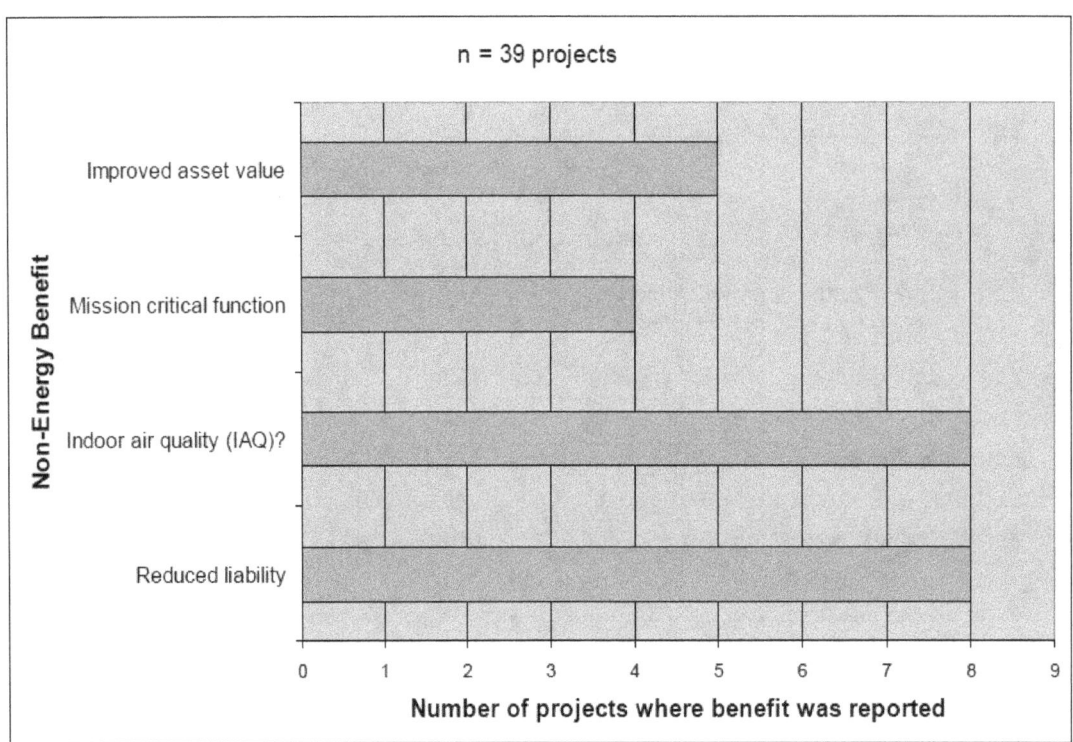

Figure 5.23 Non-energy benefits – asset value and liability reduction

There has been no attempt to quantify these benefits, but it does confirm claims that EBCx provides a number of recognizable benefits in addition to energy savings.

5.4 Discussion

Key Results

The data collected through this research project begins to characterize the various types of commissioning processes that are occurring in Annex member countries internationally. While data was often difficult to obtain, we expanded our knowledge in the following key areas:

- the scope of the Cx process employed for new and existing buildings

- characterization of issues discovered through the Cx process including system type, likely origin of issue (design, construction/installation, O&M, or capital improvement), issue type, and measures implemented

While these results begin to develop a qualitative picture for how commissioning is evolving internationally, quantitative results were less apparent. For example, data on commissioning costs and energy savings were highly variable. Falling short of the data collection goals set by Annex member country representatives at the beginning of the project, we were not able to compile a large enough sample to make strong conclusions about the cost-effectiveness of Cx internationally.

Given the relatively small sample size and variable nature of the costs and benefits, it is interesting to note that for existing building commissioning, the median savings value of 8 % corresponds well with PECI's industry experience. Further, the 3.7 year simple payback period found in this study is much higher than the LBNL 2004 study which reported a median simple payback period of 0.7 years. Industry experience has shown that typical EBCx paybacks fall in the 1year to 2 year range, which indicates that the LBNL study may not have accurately accounted for costs or counted benefits from measures that were not implemented. This research project simple payback is expected to be higher than the U.S. industry average due to the developing nature of the commissioning industry internationally, and the fact that many of the projects in the cost-benefit analysis were performed by researchers.

Data Access Limitations

One of the key barriers to obtaining a large set of international commissioning cost-benefit surveys was the limited data accessibility for Annex participants, especially for new building commissioning projects. The most difficult pieces of information to collect were data on costs and benefits, while qualitative data on the commissioning process and outcomes was more readily available. By setting a data collection methodology beforehand, we hoped to avoid retroactive data collection relying solely on documentation – "take what you can get".

Costs

The cost of commissioning was difficult to obtain because many projects were research or demonstration projects, which have costs above and beyond what would be found in typical commissioning projects. Breaking out the research costs from the commissioning costs was not a clear process and in general, these projects reported higher costs than average. In some projects, the commissioning cost by building was not separately tracked due to the contracting relationship between the provider and a large campus of buildings. In these cases, costs by building were estimated as the percent of total fee.

When the commissioning process is not performed by a third party, as often was the case in international projects, costs can be difficult to estimate as a portion of the design or construction costs and accounted for consistently. Even when commissioning was performed by a third-party whose costs were accurately reported, we attempted to account for the cost of commissioning beyond simply tracking this commissioning provider fee. For example, the cost to the controls contractor to assist the commissioning provider in functional performance testing may be viewed as a cost of commissioning. However, this level of detailed tracking was not generally submitted.

Energy Benefits

From the outset of the research project, it was known that energy cost savings data for new construction commissioning is difficult to report due to lack of baseline to compare the first year energy consumption against. New construction commissioning savings requires a methodology for assessing the baseline, typically through simulation that is calibrated to actual utility bills. Alternately, savings can be estimated through engineering calculations based on the measures found and fixed through the commissioning process. With only six buildings reporting energy savings for new construction commissioning, this points to the need for more focused research efforts in this area.

In tracking energy benefits, Cx providers do not always track by finding and often do not measure whole building energy savings as a result of EBCx projects. Owners often have less rigorous documentation requirements than projects that include a utility program incentive payment – utilities generally require detailed energy savings calculations up front prior to the decision to implement, then some kind of monitoring or verification that the measures were appropriately implemented. Contrasting the utility-sponsored EBCx project with an owner-sponsored one, an owner may simply provide a fee for "fixing my building", without requiring measure by measure analysis. For these cases, whole building energy consumption may be reviewed to determine total savings, which was the method performed for 26 of the buildings submitted.

Tracking and reporting the persistence of energy benefits from new and existing building commissioning generally requires a focused persistence study. While we included a section in the methodology for providing persistence data, the only instances in which it was used was for projects reported by Texas A&M. Since those projects have already been summarized in other research findings, the results have not been included in this report.

Non-energy benefits

Through this research, non-energy benefit data was gathered from a qualitative perspective, as it related to three areas: O&M, indoor environment, and asset value/liability. It is difficult to quantitatively assess these benefits without developing standard methods, and this process was out of scope for the cost-benefit research.

State of the Commissioning Industry Internationally

One of the results of this cost-benefit research was to develop a greater understanding of the maturity of the Cx industry internationally. Table 5.4 summarizes the phase of commissioning industry adoption for Annex 47 member countries that participated in this research. It was developed through discussion at Annex meetings and is based on findings from a recent research paper on the international commissioning industry development (Castro et al; 2008).

Table 5.4 Phase of commissioning industry adoption for Annex 47 member countries

	Research	Early adopters		Developing industry		Different framework*	
		Cx	EBCx	Cx	EBCx	Cx	EBCx
US	X			X	X		
Canada	X	X	X				
Germany*	X				X	X	
Japan	X	X	X				
Czech Republic	X						
Belgium	X	X	X				
Netherlands	X	X	X				
Norway	X	X					
Hungary	X	X	X				

* Germany follows a different framework for implementing Cx services, where Cx activities are required by code for various parties.

All Annex countries have performed Cx/EBCx research at universities and through country-sponsored research programs. Much of this research involved developing Cx guidelines specific to these countries and tools to help streamline detection of problems in buildings. A few countries can be considered "early adopters" for Cx, with a few practitioners offering Cx services as part of their service portfolio, but without any established national industry groups or guidelines.

Commissioning is developing into a recognized industry in the U.S., although Canada is moving towards this as it builds its infrastructure to deliver these services. It is interesting to note that Canada created the first guideline for the commissioning process, prior to the U.S. industry development. Existing building performance evaluation is also a well-developed field in Germany.

Germany follows a different framework for implementing new construction Cx services, where Cx is not considered as a separate activity. In Germany, 80 % to 90 % of the typical Cx tasks are regulated by (mandatory) German codes and standards. In no other country is the commissioning process so widely regulated. From the client's perspective, the quality control is part of the typical project delivery process in Germany. In the U.S., many clients hesitate to spend money for an additional Cx service, since they do not understand the value.

For example, in Germany the contractor is required not only to perform testing, but also to document the results; to adjust actuators and optimize controls; to provide training; to compile systems manuals; etc. There is always an authority overseeing and approving these activities which is usually a construction supervisor from the design firm and the tasks and responsibilities are clearly defined within the regulations. Often, this person (or team) is on-site every day or at least several days a week during the entire construction phase. In projects in the U.S., the construction supervision, that is done by the design team (e.g. mechanical engineer), typically happens every other week or once a month, and, not every project actually has a separate Cx provider on board. When we consider Cx as a quality control during construction and start-up, this is given in Germany; however, not by an independent 3rd party, and, definitely depends on the skills of the construction supervisor.

Further, in Germany there is a requirement that starts after project completion for the involvement of the design firm for usually another two years during occupancy (or whatever the warranty period of the contractors/manufacturers is). The most important tasks within this phase is a walk-through to check functionality of the equipment and to make accordant warranty claims against the contractors. Done by experienced firms, this would includes the optimization of system operation.

5.5 Recommendations

Based on the results of this research project, the following potential paths forward are recommended for future international as well as U.S. commissioning cost-benefit analyses.

International Commissioning Cost-Benefit Analysis

As the commissioning process is gaining momentum internationally, each country benefits from better definition of the value proposition. In order to obtain enough relevant data to show conclusive cost-effectiveness results for new construction and existing building commissioning in countries where the commissioning process is in its infancy, the cost-benefit methodology developed through this research project should be revisited when the commissioning industry is more mature. This project has set up standard methods that can be tailored to local needs going forward. For instance, if commissioning is integrated into the new building design and construction process and typically performed by the design team, then appropriate break-out percentages could be developed as guidance for accounting for Cx-related costs and benefits.

An additional recommendation for future international commissioning cost-benefit analysis is to obtain co-funding from participating countries to have a data collection point person that is familiar with the research collect and input the commissioning data into the survey, rather than volunteer researchers and industry contacts. Each point person would be responsible for collecting and analyzing the data from their country according to this protocol, with issues such as variances in the commissioning process well-understood by the researcher. This country-level analysis would result in a well-populated international Cx cost-benefit database where commonalities are analyzed.

US Commissioning Cost-Benefit Analysis

While the international commissioning process is continuing to develop, the U.S. should continue to expand its collection of commissioning data to meet current needs. One of the most pressing needs is to develop ways to scale up existing building commissioning efforts throughout the country to meet aggressive energy saving goals and mandates. There is a growing need for understanding the most common problems found through the EBCx process across the country, and the measures that have been implemented to address these problems. Existing cost-benefit analyses in the U.S. have focused on collection of total cost and savings information, rather than measure-level details.

Through this research project and other related work, obtaining accurate commissioning cost information for EBCx was difficult due to the variety of EBCx scope of work included (i.e., whole building commissioning, energy-focused commissioning, tune-up, etc.). While a true understanding of EBCx cost is important for owner acceptance of a rigorous process, we should not to let this hurdle derail our efforts to gather other commissioning data that is more readily available.

Due to these factors, it is recommended that future research focuses on the problems found in buildings and their solutions, whether these problems were found through a whole building or less comprehensive commissioning process. Through this process, commonly problematic systems and technologies are uncovered, where changes need to be made to decrease problems or improve training. This research would help develop focused programs targeting these commonly found problems and would lead to better guidance on how to find and fix the problems in an integrated, holistic way. Further, a solid understanding of EBCx findings would help define early solutions to avoid the occurrence of these issues in the first place.

By focusing on cataloging building problems found through EBCx, we avoid the complexities of teasing out costs of EBCx when the process is applied differently by different providers and owners. It will also be important to collect energy savings data where it exists and clearly catalogue the methodology for estimating the savings (according to the existing cost-benefit protocol).

There are still complexities with this approach to tease out, including:

- Are these the most common problems because they are the easiest to find? Are there bigger problems that are masked?

- Are bigger problems avoided because the energy savings analysis is too difficult? Is this a consequence of the energy savings calculation review process in EBCx utility incentive programs?

Finally, it is worth considering this data in the context of the drivers for market growth. In the U.S. for example, utility programs are the major drivers for growth of existing building commissioning, and USGBC's LEED certification is an emerging market driver for new construction commissioning. Engaging with these market drivers should improve the integrity/volume of data collected, and should also maximize the effectiveness of research outputs.

5.6 References

N. S. Castro, N. Nakahara, H. Yoshida, "The International State of Building System Commissioning" 2008 ACEEE Summer Study on Energy Efficiency in Buildings , Pacific Grove, CA, (17-Aug-2008) (PubID: 861505)

Friedman, H., M. Frank, T. Haasl, K. Heinemeier, 2005. "Abbreviated Commissioning Cost-Benefit Methodology."

Friedman, H., M. Frank, T. Haasl, K. Heinemeier, 2006. Chapter 1 "State-of-the-Art Review for Commissioning Low Energy Buildings"

6. CONCLUSIONS ON COMMISSIONING COST-BENEFIT AND PERSISTENCE

Commissioning Cost-Benefit

The data collected through this research project begins to characterize the various types of commissioning processes that are occurring in Annex member countries internationally. While data was often difficult to obtain, we expanded our knowledge in two key areas:

- The scope of the Cx process employed for new and existing buildings

- Characterization of issues discovered through the Cx process including system type, likely origin of issue (design, construction/installation, O&M, or capital improvement), issue type, and measures implemented

For new construction Cx, most projects included developing a Cx plan, functional testing, issue resolution, and a final report. This finding was expected since a partial commissioning process is often employed, where commissioning begins during the construction phase. The least often included items in scope (upated as-built drawing, systems/EBCx manual, and diagnostic tools) are also consistent with our understanding of the U.S. experience. These items have high relevance for improving the persistence of benefits from commissioning.

Based on this research, the scope of EBCx internationally generally includes trend analysis, master list of findings, presentation of the findings report to the owner, implementation of operations and maintenance improvements, and a final report. The low occurrence of updating documentation, development of a systems/EBCx manual, and monitoring persistence is a result of the cost of these activities. The lack of documentation and monitoring may result in a lack of persistence of benefits from these EBCx projects.

The issues found through both new construction commissioning and existing building commissioning most often occurred in air handling systems, heating water plants, and chilled water plants. In new construction projects, the origin of these issues was evenly split between design, construction/installation, and O&M/controls. For existing building projects, more issues were related to O&M/controls (42 %), with the second most common being design (31 %). Additionally in existing building commissioning projects, 27 % of issues were related to design/installation, 17 % related to setpoints, and 11 % each were related to scheduling, resets, and controls.

While these results begin to develop a qualitative picture for how commissioning is evolving internationally, quantitative results were less apparent. For example, data on commissioning costs and energy savings were highly variable. Falling short of the data collection goals set by Annex member country representatives, it was not possible to make strong conclusions about the cost-effectiveness of Cx internationally. However, progress was made towards understanding and categorizing the state of the commissioning industry for new and existing buildings in Annex member countries. While all countries have Cx research occurring, the majority of countries are in an early adopter phase of industry development. Only a few countries can be categorized as having a developing commissioning industry in which services are becoming more commonly obtained by owners.

Persistence of Commissioning Benefits

A review of results of studies from five projects related to the persistence of commissioning benefits, either in new or existing buildings conducted prior to the beginning of Annex 47 found that the savings in the buildings that were retro-commissioned generally showed some degradation with time. In retro-commissioned buildings, savings generally decreased with time, but there is wide variation from building to building.

For the new buildings, well over half of the 56 commissioning fixes persisted. Hardware fixes, such as moving a sensor or adding a valve, and control algorithm changes that were reprogrammed generally persisted. Control strategies that could easily be changed, such as occupancy schedules, reset schedules, and chiller staging tended were less likely to persist. It was also found that the extent to which persistence occurs is also related to operator training.

Persistence studies conducted as part of the Annex 47 work found that as a group, four Japanese buildings studied are consistent with the commissioned buildings examined in the literature review – most continue to show savings over periods that run from 10 years to 20 years, while generally showing some decrease in savings over time. Additional study of 10 buildings at Texas A&M found that cooling and heating savings did not change appreciably during the up to seven years of additional data following the year 2000. The average cooling savings were 45 % in 1997, 34 % in 2000 and 36 % in the last year of good data for each building (average data went through 2006) Heating savings averaged 63 % in 1997, 56 % in 2000 and 55 % in the last year of available data for each building. It should be noted that ongoing commissioning follow-up was conducted in most of these buildings at least once.

A preliminary study conducted during the Annex found that determination of savings using a "normalized annual consumption" as the basis for savings determination produced less variation in savings and persistence than found when the actual weather during the baseline and post-commissioning periods was used. It also suggested that use of calibrated simulation for baseline determination may provide more stable results.

Two prototype computerized tools that can assist in maintaining the persistence of commissioning savings have been demonstrated:

- The Automated Building Commissioning Analysis Tool (ABCAT) is a simple automated tool that uses calibrated simulation to maintain the optimal energy performance in a building. It can continuously monitor building energy consumption, alert operations personnel early upon the onset of significant increase in consumption and assist them in identifying the problem. In the six live building implementations, over eight building-years of operation, ABCAT identifed eight periods where significant energy consumption changes occurred that otherwise went undetected by the building energy management personnel. In the five retrospective building test cases (approximately 20 building years of analysis), ABCAT detected 18 faults.

- The Diagnostic Agent for Building Operation (DABO), a BEMS-assisted commissioning tool monitors the enormous amounts of data produced by BEMS and provides an extensive analysis of the incoming data. The use of a detailed systematic automated approach has improved the quality assurance process and the overall performance of the building. Furthermore, automating this essentially manual process has allowed its application on an on-going basis, generating benefits over the entire life of the CETC-Varennes building. The optimization process and the 'DABO^{TM}' tool are being demonstrated in more than ten projects. These projects include some of the first Canadian LEED buildings and the participation of major Canadian facility management firms and commissioning providers.

Current information on persistence of commissioning energy savings in existing buildings may be summarized as:

- Savings persistence at the time of the study (3 years to 20 years after commissioning) ranged from about 50 % to 100 % in all but a handful of buildings.
- Average savings at the time of the study were about 75 % of the original savings.
- The most dramatic savings degradation was caused by undetected mechanical or control component failures.
- Follow-up when needed has demonstrated persistence of commissioning savings for 7 to 20 years in a small number of buildings.

APPENDIX A: Commissioning Cost-Benefit Database

I: Commissioning Cost-Benefit Protocols Report

II. Data Collection Form (Cx)

III: Data Collection Form (EBCx)

IV: Tabulated Survey Results (Cx)

V: Tabulated Survey Results (EBCx)

IEA Annex 47:
Cost-Effective Commissioning for
Existing and Low Energy Buildings

Cost-benefit Protocols Report

*A report documenting the cost-benefit methodology and protocols
agreed to by the Annex participants*

Submitted to:
Phil Haves, LBNL

Submitted by:
PECI

May 23, 2007

1.0 INTRODUCTION

1.1 Background

This report, documenting the cost-benefit protocols and methodology agreed to by participants in the IEA's Annex 47, represents the culmination of an 18-month process in which the goals and preferences of Annex members were ascertained and a series of data collections forms were developed, tested, revised and tested again.

2005

This work began in December 2005 with the delivery of an "abbreviated methodology," based on the results of a survey of Annex members conducted at the Prague Annex meeting in October, 2005. The abbreviated methodology specified the project goals, data collection format and types of cost, energy and non-energy benefit data that Annex members wanted to collect (Friedman et al. 2005).

2006

The first draft of the data collection form, an Excel spreadsheet, was completed in March 2006 and a revised version – "v2.0," was completed in July 2006. A field test of v2.0 was conducted between July and October 2006, in which the form was distributed to approximately 1,000 recipients via email. They included Annex members, Evan Mills of Lawrence Berkeley National Laboratory and the mailing lists of the California Commissioning Collaborative, and the Building Commissioning Association.

During the field test, recipients were asked to use the form to submit data on commissioning projects. PECI received nine completed forms. Results of the field test were presented to Annex members at the Trondheim Annex meeting in April 2006 and a discussion was held to resolve a set of outstanding issues, including terminology, system types to include and normalization options.

After completion of the field test, the form was revised based on comments received and a new version – "v3.0," was presented to Annex members at the Shenzhen Annex meeting in October 2006. Another discussion was held with participants to resolve additional outstanding issues, including the extent of required data, numeral formatting, definitions of conditioned v. non-conditioned floor area, terminology and building and system types to include.

2007

In March 2007 the data collection form was revised for a final time reflecting all of the comments received to-date and the final version – "v3.1," was presented to Annex members at the Budapest Annex meeting in April 2007. In comparing the final data collection form to the initial form developed in March 2006, one will observe a significant evolution. Major changes include the use of separate forms for new and existing buildings rather than a single form for both, the division of the form into four tabs to make completion easier and expanded sections for collecting data on non-energy benefits, issues and measures.

1.2 This report

This report summarizes the data collection protocols agreed to by the Annex members. Its organization mirrors that of the data collection forms, and includes four sections: commissioning project, energy, non-energy benefits and issues and measures. The data collection forms themselves are attached as separate documents.

2.0 COMMISSIONING PROJECT

2.1 Confidentiality

Respondents are asked whether the building name or any other information on the form must remain confidential.

2.2 Project information

This section includes the name and contact information for the person completing the form, the name of the commissioning provider, the building name and location, as well as other data about the building and the project, including:

- Design temperatures
- Building type
- Occupancy type
- Performance of O&M
- Awards received
- Year commissioning completed
- Owner's primary and secondary reasons for commissioning
- Who led the process and who contracted with the commissioning provider

New building projects include the following additional data fields:

- Year construction completed
- Project delivery method
- Whether occupancy occurred on schedule

Existing building projects include the following additional data fields:

- Did the building undergo commissioning when constructed, and if so, in what phase did commissioning begin

2.3 Technical information

This section provides a technical overview of the building and the commissioning project. It is the same for new and existing buildings. Data fields include:

- Floor area units
- Total floor area and that to be used in an analysis of costs and benefits

- Number of buildings (if more than one) included in the floor area
- Number of floors and their average height
- Number of operating hours per week
- Whether the building is served by a central heating and cooling plant

This section also asks respondents to identify the number of issues found in each of the commissioned systems and to identify which percentage of issues originated in each phase of the building's history: design, construction/installation, operational/controls/maintenance, or capital improvement.

2.4 Cost data

This section aims to record the various costs of the commissioning project and the activities represented by the costs. Data fields include:

- Currency units
- Year costs incurred
- Commissioning provider free
- Who paid the costs
- Costs incurred in each phase of the process
- Items included in the cost estimates (list differs for new and existing buildings)

The new building form also asks for the building construction cost and costs to other parties to conduct commissioning tasks (architect, mechanical engineer, general contractor, mechanical contractor, electrical contractor, controls contractor and the owner's O&M staff).

The existing building form asks for slightly different costs, including the costs to other parties to conduct commissioning tasks (controls contractor and the owner's O&M staff), the cost to implement measures and the cost of monitoring and verification.

3.0 ENERGY

3.1 Source, unit and cost

This section collects the unit type, energy source and average cost per unit for electricity, electric demand, fuel, district chilled water, district hot water, district steam and water.

3.2 Resource savings

In new buildings, respondents are asked to submit the first year annual resource use for each resource type and to estimate the savings obtained through commissioning, as well as the method used to determine the savings.

In existing buildings, respondents are asked the annual resource use before commissioning for each resource type, the savings from implemented measures, the estimated savings from measures recommended but not implemented and the method used to determine the savings.

In both forms, respondents are asked whether data could have been affected by major changes in occupancy, conditioned floor area, building use and major equipment. Respondents are also asked whether the data submitted has been normalized for weather, changes in occupancy and floor area and, if so, to what year, percentage occupancy and floor area.

3.3 Resource use after commissioning

This section was developed in collaboration with David Claridge, Texas A&M University. Respondents are asked about the percent resource savings for each fuel type which continued to accrue for up to six years post-commissioning and asked to submit the starting month and year of the baseline period.

This section is to be completed only if multiple years of post-commissioning data are available. All data must be weather normalized.

4.0. NON-ENERGY BENEFITS

4.1 New buildings

This form asks for data on non-energy benefits in the following areas:

- Project cost savings
- Benefits to project design
- Change orders and warranty claims
- Construction team coordination
- Project schedule
- Building start-up and turn-over
- Operations & maintenance (O&M)
- Indoor environment

4.2 Existing buildings

This form asks for data on non-energy benefits in the following areas:

- Asset value
- Liability reduction
- Operations & maintenance (O&M)
- Indoor environment

5.0. ISSUES AND MEASURES

5.1 Issue data

Data is requested on issues (or problems) discovered during the commissioning process, including:

- Issue description (free response)
- System affected (drop-down list)
- Equipment affected (drop-down list)
- Issue category (drop-down list)
- Issue type (drop-down list)
- Origin of issue (drop-down list)

5.2   Measure data

Data is requested on measures (or solutions) recommended as a result of the issues discovered, including:

- Measure description (free response)
- Whether measure was implemented (drop-down list)
- Implementation cost (free response)
- Measure cost type (drop-down list)

5.3 Resource savings

Data is requested on the resource savings resulting from each measure and for each fuel type. Respondents are also asked to provide:

- Savings estimation approach (drop-down list)
- Total resource cost savings (free response)
- Number of years after commissioning that 50 % or more of the resource savings continued to accrue (drop-down list)
- An explanation of why the savings did or did not continue to accrue (free response)

5.4 Non-energy benefits

Data is requested on the non-energy benefits resulting from each measure. For both the primary and secondary non-energy benefits (selected from a drop-down list), respondents are also asked to provide the monetary value of the benefit and the calculation method used to determine the monetary value.

5.5 References

Hannah Friedman, Marti Frank, Tudi Haasl and Kristin Heinemeier, "Annex 47 Abbreviated Commissioning Cost-Benefit Methodology," December 9, 2005.

II. Data Collection Form (Cx)

DATA COLLECTION FORM (Cx) – "INSTRUCTIONS" SECTION

Commissioning Data Collection Form - New Buildings

IEA Annex 47   Version 4.0   September 2007

1. Data entry instructions

a. Enter data in *white cells only*.

b. Cells with red boxes are required information.

c. Leave unknown fields blank. It is not expected that all data will be available for all projects.

d. Data entry restrictions are in place. Some cells contain dropdown lists, others require specific data formats (for example, a whole number).

2. Overview of each tab

a. *Commissioning Project:* Confidentiality and contact information, data on the project's technical aspects and cost.

b. *Energy:* Source units and costs, savings resulting from the project, continued savings after the project.

c. *Non-energy benefits:* Qualitative and quantitative data on cost savings, benefits to project design, change orders and warranty claims, construction team coordination, project schedule, start-up and turnover, O&M, indoor environment.

d. *Issues and Measures:* Detailed data on issues (problems) discovered during the project and measures (corrective actions).

e. *IPMVP Definitions:* Explanation of five possible M&V options, to be used in completing the survey when prompted by a drop-down menu.

3. Submit completed forms:

Please email completed forms to: Hannah Friedman, PECI (US): hfriedman@peci.org

4. Questions?

Contact Hannah Friedman, PECI (US):   503-595-4492 or hfriedman@peci.org

Revision History: Current version is v4.0. Previous versions include v2.0 (Fall 2006) and v3.1 (Spring 2007).

Credits: Developed by PECI with comments from Annex members, based on cost-benefit methodologies from Lawrence Berkeley National Laboratory and the California Commissioning Collaborative.

DATA COLLECTION FORM (Cx) – "COMMISSIONING PROJECT" SECTION

Commissioning Data Collection Form - New Buildings

	IEA Annex 47	Version 4.0	September 2007		
		Notes	Project Data	Units	Comments
CONFIDENTIALITY					
Is it necessary for the building name to remain confidential?					
PROJECT INFORMATION					
Name of person completing form					
Contact information of person completing form: Phone					
Contact information of person completing form: E-mail					
Hours spent completing this form					
Who led the commissioning process?					
Name of commissioning leader					
How many other commissioning projects did the commissioning leader complete prior to this project?		Include only projects completed by leader, not firm			
Who held the contract with the commissioning leader?					
Name of building/project					
Building location					
City					
State					
Country					
Year construction completed					
Total building construction cost		Total cost paid by owner for the building or retrofit			

PROJECT INFORMATION CONTINUED

Building ownership			
Building occupancy type			
Who performs operations and maintenance?			
Project delivery method	Mouse over for definitions		
Commissioning project type			
In what phase did the commissioning begin?			
Year commissioning project completed			
Was commissioning undertaken in part to achieve an award or certification?			
Awards/certifications received	Select "Y" for all awards received List rating level in column E		
LEED-NC			
Energy Star (USA)			
CASBEE (Japan)			
EPBD (EU)			
Other	List award name and rating level in column E		
Indicate the owner's reasons for commissioning on a scale from 1 to 5, where "1" is very important and "5" is not important			
Ensure system performance			
Obtain energy savings			
Ensure or improve occupant comfort			
Extend equipment life			
Train and increase awareness of building operators			
Smoother process and/or turnover			
Increase occupant productivity			
Ensure adequate indoor air quality (IAQ)			

PROJECT INFORMATION CONTINUED

Comply with LEED or other sustainability rating system		
Reduce liability		
Qualify for rebate, financing, or other services		
Participate in a research, demonstration or pilot project		
Participate in a utility program		
Other	Explain in column E	
Level of commissioning	If only energy efficiency measures were commissioned, select "Specific systems"	

TECHNICAL INFORMATION

Floor area units		
Total floor area	*automatically calculated*	#N/A
Floor area served by commissioned systems (excludes parking)	*automatically calculated*	#N/A
Total floor area dedicated to each of the following uses		
School		#N/A
University		#N/A
Hospital or health facility		#N/A
Laboratory		#N/A
Office		#N/A
Hotel		#N/A
Retail		#N/A
Restaurant		#N/A
Supermarket		#N/A
Residential apartment building		#N/A
Parking		#N/A
Public assembly		#N/A
Public order and safety		#N/A
Religious worship		#N/A

TECHNICAL INFORMATION CONTINUED

Industrial building			#N/A
Service, warehouse or storage			#N/A
Vacant			#N/A
Other	Explain in column E		#N/A
If the indicated floor area is for multiple buildings, how many buildings?			
Is the facility served by a central heating and cooling plant that serves multiple buildings?			
How many issues were found in each commissioned system? If the system was not commissioned, leave cell empty. If the system was commissioned but no issues were found, enter 0. If the same issue occurred multiple times or places, only count as one issue.	Each system category below includes the controls related to the system		
HVAC system integration (EMCS/BAS)			
Chilled water plant and distribution system			
Packaged or split system DX			
Heating water plant and distribution system			
Domestic hot water			
Heat pump system			
Air handling and distribution: Overhead system			
Air handling and distribution: Underfloor system			
Terminal units			
Thermal energy storage			
Radiant heating			
Radiant cooling	Mouse over for definition		
Passive heating/cooling			
Natural ventilation or mixed-mode ventilation			
Lighting/daylighting and lighting controls			
Electrical			
Plumbing			
Envelope and infiltration			
Fire/life safety			
Utility-related (electric, gas, water, emergency power)			

TECHNICAL INFORMATION CONTINUED

Item	Notes	Value
Security		
Refrigeration		
Telecommunications		
Plug loads		
Other	Explain in column E	%
What percent of the issues above fall into each of the following categories:		
Design Issue	*automatically calculated* must equal 100 %	%
Construction and installation issue		%
Operational/controls or maintenance issue		%
Capital improvement		%
COST DATA	Give costs in year of original data; do not correct for inflation	
Currency		
Commissioning costs		
Cost unit (currency or labor hours)		
Total commissioning costs		#N/A
Commissioning leader fee	*Please provide if available* Do not include change-order costs	#N/A
Costs to other parties to performing commissioning functions	*Please provide if available* For example, if contractor writes the prefunctional checklists, include this cost. Do not include change-order costs or cost to implement measures	
Architect		#N/A
Mechanical Engineer		#N/A
General Contractor		#N/A
Mechanical Contractor		#N/A

COST DATA CONTINUED

Electrical Contractor		#N/A
Controls Contractor		#N/A
Owner's Operation and Maintenance Staff		#N/A
Percent of cost paid by:	*automatically calculated must equal 100 %*	%
Building owner		%
Utility (for example, as a rebate or incentive)		%
Grants/tax incentives		%
Other	Explain in column E	%
Percent of cost incurred in each phase	*automatically calculated must equal 100%*	%
Pre-design phase		%
Design phase		%
Construction phase		%
Warranty/occupancy phase		%
Items included in cost estimates	Select "Y" if item was included in cost estimates	
Development of owner's project requirements and basis of design (if not well-developed by designer)		
Building modeling		
Write commissioning specifications		
Develop commisioning plan		
Design review		
Develop sequences of operation (if not well-developed by mechanical or controls contractor)		
Submittal review	Mouse over for definition	
Construction observation		
Verification checks/pre-functional testing		
Use of diagnostic tools and Cx automation techniques	List tools and/or methods used in column E	
Functional testing		
Significant involvement in issue resolution		

COST DATA CONTINUED

Oversee training				
Review operations & maintenance manuals				
Develop systems manual/recommissioning manual				
Perform trend analysis				
Evaluate energy cost savings				
Final commissioning report				
Update as-built drawings				
Other	Explain in column E			

DATA COLLECTION FORM (Cx) – "ENERGY" SECTION

Commissioning Data Collection Form - New Buildings

IEA Annex 47	Version 4.0	September 2007			
	Notes	Project Data	Units	Comments	

SOURCE, UNIT & COST

	Notes	Project Data	Units	Comments
Electricity				
Unit				
Average cost per unit (in the first year after commissioning)			#N/A	
Electric demand				
Unit				
Average cost per unit (in the first year after commissioning)			#N/A	
Fuel				
Fuel type				
Unit				
Average cost per unit (in the first year after commissioning)			#N/A	
District chilled water				
Unit				
Average cost per unit (in the first year after commissioning)			#N/A	
District hot water				
Unit				
Average cost per unit (in the first year after commissioning)			#N/A	
District steam				
Unit				
Average cost per unit (in the first year after commissioning)			#N/A	
Water				
Unit				
Average cost per unit (in the first year after commissioning)			#N/A	

WHOLE BUILDING RESOURCE SAVINGS	To calculate savings, subtract first year annual usage from projected usage of baseline building (the original design or building, before commissioning began)	
Electric consumption		
First year annual usage		#N/A
Estimated annual savings from commissioning		#N/A
Describe method used to determine savings and baseline		
Electric demand		
First year peak demand		#N/A
Estimated demand reduction from commissioning		#N/A
Describe method used to determine savings and baseline		
Fuel		
First year annual usage		#N/A
Estimated annual savings from commissioning		#N/A
Describe method used to determine savings and baseline		
District chilled water		
First year annual usage		#N/A
Estimated annual savings from commissioning		#N/A
Describe method used to determine savings and baseline		
District hot water		
First year annual usage		#N/A
Estimated annual savings from commissioning		#N/A
Describe method used to determine savings and baseline		
District steam		
First year annual usage		#N/A
Estimated annual savings from commissioning		#N/A
Describe method used to determine savings and baseline		
Water		
First year annual usage		#N/A

WHOLE BUILDING RESOURCE SAVINGS CONTINUED

Estimated annual savings from commissioning		#N/A
Describe method used to determine savings and baseline		
Could data have been affected by:		
Major occupancy changes?		
Average % occupied **before** commissioning		%
Average % occupied **after** commissioning		%
Changes in conditioned floor area?		
Average % conditioned floor area **before** commissioning		%
Average % conditioned floor area **after** commissioning		%
Changes in building use	If yes, describe in column E	
Major equipment changes	If yes, describe in column E	
Has submitted data been normalized for:		
Weather?		
If yes, to what year was the data normalized?		
Changes in occupancy?		
If yes, to what % occupancy was the data normalized?		%
Changes in floor area?		
If yes, to what floor area was the data normalized?	If yes, describe in column E	#N/A
Other	If yes, describe in column E	
RESOURCE USE AFTER COMMISSIONING	Please complete this section if multiple years of post-commissioning use data are available	
Indicate the "baseline" period (must be one full year of data - usually the first full year after commissioning)	Example: enter January as 1	
Start month		
Start year		
End date	automatically calculated	12/1900

Electric consumption - weather normalized % savings from baseline

Year 1		%
Year 2		%
Year 3		%
Year 4		%
Year 5		%
Year 6		%

Electric demand - weather normalized % savings from baseline

Year 1		%
Year 2		%
Year 3		%
Year 4		%
Year 5		%
Year 6		%

Fuel - weather normalized % savings from baseline

Year 1		%
Year 2		%
Year 3		%
Year 4		%
Year 5		%
Year 6		%

District hot water - weather normalized % savings from baseline

Year 1		%
Year 2		%
Year 3		%
Year 4		%
Year 5		%
Year 6		%

RESOURCE USE AFTER COMMISSIONING CONTINUED

District chilled water - weather normalized % savings from baseline				
Year 1	%			
Year 2	%			
Year 3	%			
Year 4	%			
Year 5	%			
Year 6	%			
District steam - weather normalized % savings from baseline				
Year 1	%			
Year 2	%			
Year 3	%			
Year 4	%			
Year 5	%			
Year 6	%			
Water - weather normalized % savings from baseline				
Year 1	%			
Year 2	%			
Year 3	%			
Year 4	%			
Year 5	%			
Year 6	%			
Has submitted data been normalized for any changes in:				
Occupancy?				
Schedules?				
Equipment?				
Occupied floor area?				

DATA COLLECTION FORM (Cx) – "NON-ENERGY BENEFITS" SECTION

Commissioning Data Collection Form – New Buildings

IEA Annex 47 Version 4.0 September 2007

	Notes	Project Data	Units	Comments
NON-ENERGY BENEFITS	Describe calculations in column E			
Project cost savings				
Did you achieve an overall lower project cost (or "first cost") as a result of commissioning?				
What is the total value of this savings?	describe in column E		#N/A	
Benefits to project design				
Were improvements to system design made as a result of commissioning?				
Was equipment sized correctly as a result of commissioning?				
What is the monetary value of these benefits?	describe in column E		#N/A	
Change orders and warranty claims				
Did you reduce or avoid change orders as a result of commissioning?				
How many change orders were avoided?				
Did you reduce or avoid warranty claims as a result of commissioning?				
What is the monetary value of these benefits?	describe in column E		#N/A	
Asset value				
Was asset value improved as a result of commissioning?				
What is the total value of this benefit?	describe in column E		#N/A	
Liability reduction				
Was liability reduced as a result of commissioning?				
Did this include reductions in liability related to:				
Indoor air quality (IAQ)?				
Fire/life safety?				
Security?				
The building's mission critical function?				
NON-ENERGY BENEFITS CONTINUED				

	describe in column E	#N/A
What is the monetary value of these benefits?	describe in column E	#N/A

Construction team coordination

Was coordination among team members improved as a result of commissioning?		
Were disagreements among contractors reduced or resolved more quickly as a result of commissioning?		
What is the monetary value of these benefits?	describe in column E	#N/A

Project schedule

Did the project progress according to schedule as a result of commissioning?		
Were potential problems detected and corrected earlier than they would have been without commissioning?		
What is the monetary value of these benefits?	describe in column E	#N/A

Building start-up and turn-over

Was the building occupied on schedule as a result of commissioning?		
Were contractor call-backs reduced as a result of commissioning?		
Were testing and balancing (TAB) costs reduced as a result of commissioning?		
What is the monetary value of these benefits?	describe in column E	#N/A

Operations & maintenance (O&M)

Was system documentation available to O&M staff as a result of commissioning?		
Were O&M staff provided with training as a result of commissioning?		
Do O&M staff report satisfaction with their ability to operate and maintain the building as a result of commissioning?		
Was equipment life extended as a result of commissioning?		
Were ongoing operations and maintenace costs reduced a result of commissioning?		

NON-ENERGY BENEFITS CONTINUED

What is the monetary value of these benefits?	describe in column E	#N/A	
Indoor environment			
Was indoor air quality improved as a result of commissioning?			
Was occupant comfort improved as a result of commissioning?			
Did this include:			
Improvements to heating?			
Improvements to cooling?			
Improvements to humidity?			
Improvements to air balance?			
Was occupant productivity improved as a result of commissioning?			
Was occupant safety improved as a result of commissioning?			
What is the monetary value of these benefits?	describe in column E	#N/A	
Other benefits			
Please list any additional benefits that resulted from commissioning in column C, and describe the benefit in column E			
What is the monetary value of these benefits?	describe in column E	#N/A	

DATA COLLECTION FORM (Cx) – "ISSUES AND MEASURES" SECTION

Commissioning Data Collection Form

IEA Annex 47   Version 4.0   September 2007

ISSUE DATA: the problem

ISSUE TYPE	ISSUE TYPE *EXAMPLE*	ISSUE DESCRIPTION	ORIGIN OF ISSUE	SYSTEM AFFECTED	EQUIPMENT AFFECTED
Select from drop-down list	Example populates based on selected ISSUE TYPE	Provide a short description of the problem	Select from drop-down list	Select from drop-down list	Select from drop-down list List populates based on selected SYSTEM AFFECTED
1					
2					

MEASURE DATA: the solution

ISSUE TYPE	MEASURE TYPE	MEASURE DESCRIPTION	WAS MEASURE IMPLEMENTED ?	IMPLEMENT-ATION COST	MEASURE COST TYPE
Select from drop-down list	Select from drop-down list	Provide a short description of the solution recommended to correct the problem	Select from drop-down list If recommended but not implemented, select "No"	#N/A	Select from drop-down list
1					
2					

RESOURCE SAVINGS: the amount of reduced resource use as a result of the measure

ISSUE TYPE	ELECTRICITY	PEAK ELECTRICAL DEMAND	FUEL	CHILLED WATER	HOT WATER	STEAM	WATER
Select from drop-down list	#N/A	#N/A	#N/A	#N/A	#N/A	#N/A	#N/A
1							
2							

RESOURCE SAVINGS: the amount of reduced resource use as a result of the measure

ISSUE TYPE	SAVINGS ESTIMATION APPROACH	TOTAL RESOURCE COST SAVINGS	FOR HOW MANY YEARS AFTER COMMISSIONING DID 80% OR MORE of the RESOURCE SAVINGS CONTINUE?	WHY DID OR DID NOT RESOURCE SAVINGS CONTINUE?
Select from drop-down list	Select from drop-down list Definitions on IPMVP tab	#N/A	Select from drop-down list	Explain
1				
2				

NON-ENERGY BENEFITS

ISSUE TYPE	PRIMARY BENEFIT: TYPE	PRIMARY BENEFIT: MONETARY VALUE	PRIMARY BENEFIT: CALCULATION METHOD	SECONDARY BENEFIT: TYPE	SECONDARY BENEFIT: MONETARY VALUE	SECONDARY BENEFIT: CALCULATION METHOD
Select from drop-down list	Select from drop-down list	#N/A	Describe method used to calculate value of benefit	Select benefit type from drop-down menu	#N/A	Describe method used to calculate value of benefit
1						
2						

III: Data Collection Form (EBCx)

EBCx DATA COLLECTION FORM – "INSTRUCTIONS" SECTION

Commissioning Data Collection Form - Existing Buildings

IEA Annex 47   Version 4.0   September 2007

1. Data entry instructions

  a. Enter data in *white cells only.*

  b. Cells with red boxes are required information.

  c. Leave unknown fields blank. It is not expected that all data will be available for all projects.

  d. Data entry restrictions are in place. Some cells contain dropdown lists, others require specific data formats (for example, a whole number).

2. Overview of each tab

  a. *Commissioning Project:* Confidentiality and contact information, data on the project's technical aspects and cost.

  b. *Energy:* Source units and costs, savings resulting from the project, continued savings after the project.

  c. *Non-energy benefits:* Qualitative and quantitative data on benefits to O&M, indoor environment, asset value, liability reduction.

  d. *Issues and Measures:* Detailed data on issues (problems) discovered during the project and measures (corrective actions).

  e. *IPMVP Definitions:* Explanation of five possible M&V options, to be used in completing the survey when prompted by a drop-down menu.

3. Submit completed forms:

  Please email completed forms to: Hannah Friedman, PECI (US): hfriedman@peci.org

4. Questions?

  Contact Hannah Friedman, PECI (US): 503-595-4492 or hfriedman@peci.org

Revision History: Current version is v4.0. Previous versions include v2.0 (Fall 2006) and v3.1 (Spring 2007).

Credits: Developed by PECI with comments from Annex members, based on cost-benefit methodologies from Lawrence Berkeley National Laboratory and the California Commissioning Collaborative.

Commissioning Data Collection Form – Existing Buildings

	IEA Annex 47	Version 4.0	September 2007		
		Notes	Project Data	Units	Comments
CONFIDENTIALITY					
Is it necessary for the building name to remain confidential?					
PROJECT INFORMATION					
Name of person completing form					
Contact information of person completing form: Phone					
Contact information of person completing form: E-mail					
Hours spent completing this form					
Who led the commissioning process?					
Name of commissioning leader					
How many other commissioning projects did the commissioning leader complete prior to this project?		Include only projects completed by leader, not firm			
Name of building/project					
Building location					
City					
State					
Country					
Building ownership					
Building type					
Year built					

PROJECT INFORMATION CONTINUED

Year commissioning project completed			
Building occupancy type			
Who performs operations and maintenance?			
Was commissioning undertaken in part to achieve an award or certification?			
Awards/certifications received	Select "Y" for all awards received List rating level in column E		
LEED-NC			
LEED-EB			
Energy Star (USA)			
CASBEE (Japan)			
EPBD (EU)			
Other	List award name and rating level in column E		
Indicate the owner's reasons for commissioning on a scale from 1 to 5, where "1" is very important and "5" is not important			
Ensure system performance			
Obtain energy savings			
Ensure or improve occupant comfort			
Extend equipment life			
Train and increase awareness of building operators			
Increase occupant productivity			
Ensure adequate indoor air quality (IAQ)			
Comply with LEED or other sustainability rating ystem			
Reduce liability			
Qualify for rebate, financing, or other services			
Participate in a research, demonstration or pilot project			

PROJECT INFORMATION CONTINUED

Participate in a utility program			
Other	Explain in column E		
Level of commissioning	If only energy efficiency measures were commissioned, select "Specific systems"		
Did the building undergo a *new building* commissioning process?			
If yes, in what phase did commissioning begin?			

Floor area units			
Total floor area	*automatically calculated*	#N/A	
Floor area served by commissioned systems (excludes parking)	*automatically calculated*	#N/A	
Total floor area dedicated to each of the following uses			
School		#N/A	
University		#N/A	
Hospital or health facility		#N/A	
Laboratory		#N/A	
Office		#N/A	
Hotel		#N/A	
Retail		#N/A	
Restaurant		#N/A	
Supermarket		#N/A	
Residential apartment building		#N/A	
Parking		#N/A	
Public assembly		#N/A	
Public order and safety		#N/A	
Religious worship		#N/A	
Industrial building		#N/A	
Service, warehouse or storage		#N/A	

TECHNICAL INFORMATION CONTINUED

Vacant		#N/A	

Other	Explain in column E	#N/A	
If the indicated floor area is for multiple buildings, how many buildings?			
Is the facility served by a central heating and cooling plant that serves multiple buildings?			
How many issues were found in each commissioned system? If the system was not commissioned, leave cell empty. If the system was commissioned but no issues were found, enter 0. If the same issue occurred multiple times or places, only count as one issue.	Each system category below includes the controls related to the system		
HVAC system integration (EMCS/BAS)			
Chilled water plant and distribution system			
Packaged or split system DX			
Heating water plant and distribution system			
Domestic hot water			
Heat pump system			
Air handling and distribution: Overhead system			
Air handling and distribution: Underfloor system			
Terminal units			
Thermal energy storage			
Radiant heating			
Radiant cooling			
Passive heating/cooling	Mouse over for definition		
Natural ventilation or mixed-mode ventilation			
Lighting/daylighting and lighting controls			
Electrical			
Plumbing			
Envelope and infiltration			
Fire/life safety			
Utility-related (electric, gas, water, emergency power)			

TECHNICAL INFORMATION CONTINUED

Security		
Refrigeration		

Telecommunications		
Plug loads		
Other	Explain in column E	
What percent of the issues above fall into each of the following categories:	*automatically calculated* must equal 100%	
Design Issue	%	
Construction and installation issue	%	
Operational/controls or maintenance issue	%	
Capital improvement	%	
COST DATA	Give costs in year of original data; do not correct for inflation	
Currency		
Commissioning costs		
Cost unit (currency or labor hours)		
Total commissioning costs		#N/A
Commissioning leader fee/investigation costs	*Please provide if available* Do not include leader's cost to assist with implementation Do not include change-order costs	#N/A
Costs for other parties to perform commissioning investigation functions	*Please provide if available* Do not include costs to implement measures	
Controls contractor		#N/A
Owner's operation and maintenance staff		#N/A

| **COST DATA CONTINUED** | | |
| Cost to implement measures (include cost of all labor and materials) | *Please provide if available* | #N/A |

	Please provide if available	#N/A
Cost for monitoring and verification of savings		
Percent of cost paid by:	automatically calculated must equal 100%	%
Building owner		%
Utility (for example, as a rebate or incentive)		%
Grants/tax incentives		%
Other	Explain in column E	%
Tasks performed	Select "Y" if task was performed	
Document owner's project requirements		
Develop Commissioning Plan		
Utility bill analysis		
Benchmarking		
Trend analysis (for example, EMCS, data logging, etc)		
Building energy modeling/simulation		
Use of diagnostic tools and Cx automation techniques	List tools and/or methods used in column E	
Document master list of findings		
Calculate energy cost savings for findings		
Present a findings and recommendations report		
Update system documentation after implementation		
Implement operations and maintenance (O&M) improvements		
Implement capital improvements		
Monitor and verify energy savings		
Monitor implemented measures for persistence of benefits		
Develop systems manual/recommissioning manual		
Final commissioning report		
Other	Explain in column E	

Commissioning Data Collection Form – Existing Buildings

IEA Annex 47	Version 4.0	September 2007		
	Notes	Project Data	Units	Comments
SOURCE, UNIT & COST				
Electricity				
Unit				
Average cost per unit (in the first year after commissioning)			#N/A	
Electric demand				
Unit				
Average cost per unit (in the first year after commissioning)			#N/A	
Fuel				
Fuel type				
Unit				
Average cost per unit (in the first year after commissioning)			#N/A	
District chilled water				
Unit				
Average cost per unit (in the first year after commissioning)			#N/A	
District hot water				
Unit				
Average cost per unit (in the first year after commissioning)			#N/A	
District steam				
Unit				
Average cost per unit (in the first year after commissioning)			#N/A	

RESOURCE SAVINGS

Water

Unit		
Average cost per unit (in the first year after commissioning)		#N/A

Electricity

Annual usage before commissioning		#N/A
Annual savings from implemented measures		#N/A
Additional annual savings from measures recommended *but not implemented*		#N/A
Method used to determine savings	Definitions on IPMVP tab	

Electric demand

Peak demand before commissioning		#N/A
Demand reduction from implemented measures		#N/A
Additional annual savings from measures recommended *but not implemented*		#N/A
Method used to determine savings	Definitions on IPMVP tab	

Fuel

Annual usage before commissioning		#N/A
Annual savings from implemented measures		#N/A
Additional annual savings from measures recommended *but not implemented*		#N/A
Method used to determine savings	Definitions on IPMVP tab	

District chilled water

Annual usage before commissioning		#N/A
Annual savings from implemented measures		#N/A
Additional annual savings from measures recommended *but not implemented*		#N/A

RESOURCE SAVINGS CONTINUED

Method used to determine savings	Definitions on IPMVP tab	
District hot water		
Annual usage before commissioning		#N/A
Annual savings from implemented measures		#N/A
Additional annual savings from measures recommended *but not implemented*		#N/A
Method used to determine savings	Definitions on IPMVP tab	
District steam		
Annual usage before commissioning		#N/A
Annual savings from implemented measures		#N/A
Additional annual savings from measures recommended *but not implemented*		#N/A
Method used to determine savings	Definitions on IPMVP tab	
Water		
Annual usage before commissioning		#N/A
Annual savings from implemented measures		#N/A
Additional annual savings from measures recommended *but not implemented*		#N/A
Method used to determine savings	Definitions on IPMVP tab	
Could data have been affected by:		
Major occupancy changes?		
Average % occupied before commissioning		%
Average % occupied after commissioning		%
Changes in conditioned floor area?		
Average % conditioned floor area before commissioning		%
Average % conditioned floor area after commissioning		%
Changes in building use	If **yes**, describe in column E	
Major equipment changes	If **yes**, describe in column E	

RESOURCE SAVINGS CONTINUED

Has submitted data been normalized for:

Weather?		
If yes, to what year was the data normalized?		
Changes in occupancy?		
If yes, to what % occupancy was the data normalized?	%	
Changes in floor area?		
If yes, to what floor area was the data normalized?	#N/A	
Other	**If yes**, describe in column E	

RESOURCE USE AFTER COMMISSIONING

Please complete this section if multiple years of post-commissioning use data are available

Indicate the "baseline" period (must be one full year of data - usually the first full year after commissioning)		
Start month	Example: enter January as 1	
Start year		
End date	*automatically calculated*	12/1900

Electric consumption - weather normalized % savings from baseline

Year 1	%
Year 2	%
Year 3	%
Year 4	%
Year 5	%
Year 6	%

Electric demand - weather normalized % savings from baseline

Year 1	%
Year 2	%
Year 3	%

RESOURCE USE AFTER COMMISSIONING CONTINUED

Year 4	%
Year 5	%
Year 6	%

Fuel - weather normalized % savings from baseline

Year 1	%
Year 2	%
Year 3	%
Year 4	%
Year 5	%
Year 6	%

District hot water - weather normalized % savings from baseline

Year 1	%
Year 2	%
Year 3	%
Year 4	%
Year 5	%
Year 6	%

District chilled water - weather normalized % savings from baseline

Year 1	%
Year 2	%
Year 3	%
Year 4	%
Year 5	%
Year 6	%

District steam - weather normalized % savings from baseline

Year 1	%
Year 2	%

RESOURCE USE AFTER COMMISSIONING CONTINUED

Year 3	%
Year 4	%
Year 5	%
Year 6	%

Water - weather normalized % savings from baseline		
Year 1	%	
Year 2	%	
Year 3	%	
Year 4	%	
Year 5	%	
Year 6	%	
Has submitted data been normalized for any changes in:		
Occupancy?		
Schedules?		
Equipment?		
Occupied floor area?		

Commissioning Data Collection Form – Existing Buildings

IEA Annex 47	Version 4.0	September 2007		
	Notes	Project Data	Units	Comments
NON-ENERGY BENEFITS	Descr be calculations in column E			
Operations & maintenance (O&M)				
Was improved system documentation available to O&M staff as a result of commissioning?				
Were O&M staff provided with training as a result of commissioning?				
Do O&M staff report increased ability to operate and maintain the building as a result of commissioning?				
Was equipment life extended as a result of commissioning?				
Were unexpected equipment failures likely prevented as a result of commissioning?				
Were ongoing operations and maintenance costs reduced a result of commissioning?				
What is the monetary value of these benefits?	describe in column E		#N/A	
Indoor environment				
Was indoor air quality improved as a result of commissioning?				
Was occupant comfort improved as a result of commissioning?				
Did this include:				
Improvements to heating?				
Improvements to cooling?				
Improvements to humidity?				
Improvements to air balance?				
Was occupant productivity improved as a result of commissioning?				
Was occupant safety improved as a result of commissioning?				
What is the monetary value of these benefits?	describe in column E		#N/A	

NON-ENERGY BENEFITS CONTINUED

Asset value		
Was asset value improved as a result of commissioning?		
What is the total value of this benefit?	describe in column E	#N/A

Liability reduction		
Was liability reduced as a result of commissioning?		
Did this include reductions in liability related to:		
Indoor air quality (IAQ)?		
Fire/life safety?		
Security?		
The building's mission critical function?		
What is the monetary value of these benefits?	describe in column E	#N/A

Other benefits		
Please list any additional benefits that resulted from commissioning in column C, and describe the benefit in column E		
What is the monetary value of these benefits?	describe in column E	#N/A

EBCx DATA COLLECTION FORM – "ISSUES AND MEASURES" SECTION

Commissioning Data Collection Form

IEA Annex 47   Version 4.0   September 2007

ISSUE DATA: the problem

ISSUE TYPE	ISSUE TYPE *EXAMPLE*	ISSUE DESCRIPTION	ORIGIN OF ISSUE	SYSTEM AFFECTED	EQUIPMENT AFFECTED
Select from drop-down list	Example populates based on selected ISSUE TYPE	Provide a short description of the problem	Select from drop-down list	Select from drop-down list	Select from drop-down list List populates based on selected SYSTEM AFFECTED
1					
2					

	ISSUE TYPE	MEASURE TYPE	MEASURE DESCRIPTION	WAS MEASURE IMPLEMENTED ?	IMPLEMENTAT ION COST	MEASURE COST TYPE
	Select from drop-down list	Select from drop-down list	Provide a short description of the solution recommended to correct the problem	Select from drop-down list If recommended but not implemented, select "No"	#N/A	Select from drop-down list
1						
2						

MEASURE DATA: the solution

ISSUE TYPE	ELECTRICITY	PEAK ELECTRICAL DEMAND	FUEL	CHILLED WATER	HOT WATER	STEAM	WATER
Select from drop-down list	#N/A	#N/A	#N/A	#N/A	#N/A	#N/A	#N/A
1							
2							

RESOURCE SAVINGS: the amount of reduced resource use as a result of the measure

RESOURCE SAVINGS: the amount of reduced resource use as a result of the measure

ISSUE TYPE	SAVINGS ESTIMATION APPROACH	TOTAL RESOURCE COST SAVINGS	FOR HOW MANY YEARS AFTER COMMISSIONING DID 80% OR MORE of the RESOURCE SAVINGS CONTINUE?	WHY DID OR DID NOT RESOURCE SAVINGS CONTINUE?
Select from drop-down list	Select from drop-down list Definitions on IPMVP tab	#N/A	Select from drop-down list	Explain
1				
2				

NON-ENERGY BENEFITS

ISSUE TYPE	PRIMARY BENEFIT: TYPE	PRIMARY BENEFIT: MONETARY VALUE	PRIMARY BENEFIT: CALCULATION METHOD	SECONDARY BENEFIT: TYPE	SECONDARY BENEFIT: MONETARY VALUE	SECONDARY BENEFIT: CALCULATION METHOD
Select from drop-down list	Select from drop-down list	#N/A	Describe method used to calculate value of benefit	Select benefit type from drop-down menu	#N/A	Describe method used to calculate value of benefit
1						
2						

IV: Tabulated Survey Results (Cx)

TABULATED SURVEY RESULTS (Cx) – "Cx PROJECT" SECTION

Project ref number	PROJECT INFORMATION				Year commissioning project completed	TECHNICAL INFORMATION	Floor area units	How many issues were found in each commissioned system? If the system was not commissioned, leave cell empty. If the system was commissioned but no issues were found, enter 0. If the same issue occurred multiple times or places, only count as one issue.	HVAC system integration (EMCS/BAS)	Chilled water plant and distribution system
	Building location	City	State	Country						
1		Stuttgart	Baden-Württemberg	Germany	2004		m2			
2		Zwolle		Netherlands	2007		m2			1
3		Davenport Iowa	Iowa	USA	2006		ft2			
4		Pittsburgh	PA	USA	2004		ft2			
5		Tachikawa	Tokyo	Japan	2005		m2		4	4
6		Oita	Oita	Japan	2000		m2			1
7		Okazaki	Aichi	Japan			m2			
8		Yokohama	Kanagawa	Japan			m2			1
9		1-5-1 Higasi Shinbashi, Minatoku	Tokyo	Japan	2006		m2			
10		kobei	Wakayama	Japan	2006		m2			

TECHNICAL INFORMATION CONTINUED

Project ref number	Packaged or split system DX	Heating or water plant and distribution system	Domestic hot water	Heat pump system	Air handling and distribution: Overhead system	Terminal units	Thermal energy storage	Lighting/daylighting and lighting controls	Envelope and infiltration	Other
1										
2		1				1	1			
3	1	3			12					7
4										
5		6					2			
6							1			
7			1	1						
8		1	1	1	1		1	1	1	
9					1				1	
10										

Project ref number	COST DATA	Currency	Cost unit (currency or labor hours)	Total commissioning costs	Items included in cost estimates	Development of owner's project requirements and basis of design (if not well-developed by designer)	Write commissioning specifications	Develop commisioning plan	Design review	Develop sequences of operation (if not well-developed by mechanical or controls contractor)	Submittal review	Construction observation
1		EUR (euro)	Currency	175000		Y		Y				
2		EUR (euro)	Currency	15000			Y		Y	Y	Y	
3		USD (US dollar)	Currency	15000		Y		Y	Y		Y	Y
4		USD (US dollar)	Currency	18000				Y		Y		
5		JPY (Japanese yen)	Labor	1104		Y	Y	Y	Y	Y	Y	Y
6		JPY (Japanese yen)	Currency	30,000,000.00								
7		JPY (Japanese yen)										
8		JPY (Japanese yen)	Labor	432								
9		JPY (Japanese yen)	Currency	36,700,000.00								
10				1000		y		y	y			

Project ref number	COST DATA CONTINUED	Verification checks/pre-functional testing	Use of diagnostic tools and Cx automation techniques	Functional testing	Significant involvement in issue resolution	Oversee training	Review operations & maintenance manuals	Develop systems manual/recommissioning manual	Perform trend analysis	Evaluate energy cost savings	Final commissioning report	Update as-built drawings
1					Y				Y		Y	
2				Y							Y	
3		Y		Y	Y	Y	Y				Y	
4		Y		Y	Y	Y	Y				Y	
5		Y		Y	Y	Y	Y	Y	Y		Y	
6												
7												
8			Y			Y						
9					Y				Y	Y	Y	
10		y		y	y				y	y	y	y

TABULATED SURVEY RESULTS (Cx) – "Cx ENERGY" SECTION

Project Ref number	Country	SOURCE, UNIT & COST	Electricity	Unit	Electric demand	Unit	Fuel	Fuel type	Unit	District chilled water	Unit	District steam	Unit	Water	Unit	Average cost per unit (in the first year after commissioning)
1	Germany			Kilowatt-hour (kWh)		Kilowatt (kW)		Natural gas	Cubic meters gas (m3)							
2	Netherlands															
3	USA			Kilowatt-hour (kWh)		Kilowatt (kW)		Natural gas	Hundred cubic feet (CCF)							
4	USA			Kilowatt-hour (kWh)		Kilowatt (kW)		Not used at his facility							Gallons	
5	Japan			Kilowatt-hour (kWh)		Kilowatt (kW)		Not used at his facility							Liters	0.37
6	Japan			Kilowatt-hour (kWh)		Kilowatt (kW)		Not used at his facility							Liters	
7	Japan			Kilowatt-hour (kWh)		Kilowatt (kW)		Not used at his facility							Liters	
8	Japan			Kilowatt-hour (kWh)		Kilowatt (kW)		Not used at his facility			Megajoule (MJ)		Megajoule (MJ)		Liters	0.37
9	Japan			Megajoule (MJ)		Kilowatt (kW)									Liters	
10	Japan															

V: Tabulated Survey Results (EBCx)

Project Ref Number	PROJECT INFORMATION — Name of person completing form	Contact information of person completing form: Phone	Contact information of person completing form: E-mail	Building location — City	State	Country	Building ownership	Building type	Year built	Year commissioning project completed
1	Alexis Versele			Oud Borgerhout	Antwerp	Belgium		Other (explain in column E)	1930	
2	Alexis Versele			Wondelgem	Oost-Vlaanderen	Belgium		School	1977	
3	Alexis Versele			Wondelgem	Oost-Vlaanderen	Belgium		School	1985	
4	Stijn Van den Broecke			Mechelen	province Antwerpen	Belgium	Private	Office	2006	2008
5	Prof. M. De Paepe			Gent	Flanders	Belgium	Public	Laboratory	2003	2007
6	Hilde Breesch			Gent	Oost-Vlaandere	Belgium	Public	School		2008
7	André Chalifour			Montréal	Québec	Canada		Other (explain in column E)	1983	2003
8	Daniel Choiniere			Varennes	Québec	Canada	Public	Other (explain in column E)	1991	2006
9	Shigehiro Ichinose			Nagoya	Aichi	Japan		Office	1971	2007
10	Kazuhiro Nakazawa			Osaka	Osaka	Japan	Private	Other (explain in column E)	2005	2007
11	Katsuhiro KAMITANI			nihonbashi,chuo-ku	Tokyo	Japan	Private	Office	1968	2004
12	Katsuhiro KAMITANI			sinkawa,chuo-ku	Tokyo	Japan	Private	Office	1988	1999
13	Masahiro Shinozaki			Oita	Oita	Japan	Private	Office	1997	2007
14	Hirobumi UEDA			Nishiku	Osaka	Japan	Private	Office	1996	2008
15	Henk Peitsman			Amsterdam	n.a.	Netherlands		Office	1999	2005
16	Henk C. Peitsman			Rotterdam	n.a.	Netherlands	Private	Office	1991	1996
17	Henk C. Peitsman			The Hague		Netherlands	Private	Office		2005
18	Henk C. Peitsman			Maastricht		Netherlands	Private	University	2005	2007
19	Henk C. Peitsman			Driebergen	n.a.	Netherlands	Public	Office	1989	2008
20	Henk C. Peitsman			Hoevelaken	n.a.	Netherlands	Private	Office	1969	2008
21	Henk C. Peitsman			Amsterdam	n.a.	Netherlands	Private	Office	2002	2006

Project Ref Number	PROJECT INFORMATION	Name of person completing form	Contact information of person completing form: Phone	Contact information of person completing form: E-mail	Building location	City	State	Country	Building ownership	Building type	Year built	Year commissioning project completed
22		Henk C. Peitsman				Delft	n.a.	Netherlands	Public	Office	2002	2008
23		Henk Peitsman				The Hague	n.a.	Netherlands	Private	Office	1969	2008
24		Natasa Djuric				Trondheim		Norway		in column E)	1958	2005
25		Natasa Djuric				South Norway, Cost			Norway	Private	Hotel	2002	2005
26		Natasa Djuric				West Norway, Cost			Norway	Private	Hotel	1987	2005
27		Natasa Djuric				South Norway, Cost			Norway	Private	Hotel	2000	2005
28		Natasa Djuric				East Norway, Inland			Norway	Private	Hotel	1992	2005
29		Omer Akin				Pittsburgh	PA	USA		University	2004	2005
30		Ken Engan				College Station	TX	USA		University	1950	1997
31		Marti Frank				La Mesa	California	USA	Private	Office	1983	2001
32		Hannah Friedman				Portland	OR	USA	Public	Public assembly	1997	2005
33		Cory Toole				College Station	TX	USA	Public	University	1978	1997
34		Cory Toole				College Station	TX	USA	Public	University	1973	1997
35		Cory Toole				College Station	TX	USA	Public	University	1955	1997
36		Cory Toole				College Station	TX	USA	Public	University	1973	1996
37		Ken Engan				College Station	TX	USA	Public	University	1977	1999
38		Cory Toole				College Station	TX	USA	Public	University	1978	1996
39		Cory Toole				College Station	TX	USA	Public	University		1997
40		Cory Toole				College Station	TX	USA	Public	University		1997
41		Cory Toole				College Station	TX	USA	Public	University		1996
42		Cory Toole				College Station	TX	USA	Public	University	1995	1997
43		Cory Toole				College Station	TX	USA	Public	University	1971	1997

Project Ref Number	Floor area units	Total floor area	# issues found in each Cx'd system? - If no Cx, leave cell empty. - If Cx but no issues were found, enter 0. - If same issue occurred multiple times or places, count as 1 issue. System categories below include controls related to the system	HVAC system integration (EMCS/BAS)	Chilled water plant and distribution system	Heating water plant and distribution system	Domestic hot water	Heat pump system	Air handling and distribution: Overhead system	Air handling and distribution: Underfloor system	Terminal units	Thermal energy storage	Radiant heating	Radiant cooling	Passive heating/cooling
1	m2	1149													
2	m2	1524	6	Y			Y						Y		
3	m2	1512	6	Y			Y						Y		
4	m2	1636	7	Y	Y			Y	Y			Y	Y	Y	0
5	m2	12900	4						Y		Y				
6	m2	3472	5	Y									Y		
7	ft2	1431500	7	Y	Y	Y	Y		Y						
8	m2	4189													
9	m2	9446													
10	m2	106363	2	Y	Y	Y		0				Y			
11	m2		6	Y	Y	Y		Y				Y			
12	m2	5400	6	Y	Y	Y		Y				Y			
13	m2	29869	2		Y	Y									
14	m2	41000	4		Y	Y	Y								
15	m2	120000													
16	m2	100000	4		Y				Y	Y	Y				
17	m2	90000	4		Y	Y			Y		Y				
18	m2	20000	3	Y	Y	Y									
19	m2	18000	2	Y	0	0			Y						
20	m2	20000	6	Y	Y	Y			Y		Y				
21	m2	31000	4	Y	Y	0			Y		Y				
22	m2	10000	3	Y	Y				Y		Y				

Project Ref Number	TECHNICAL INFORMATION	Floor area units	Total floor area	# issues found in each Cx'd system? - If no Cx, leave cell empty. - If Cx but no issues were found, enter 0. - If same issue occurred multiple times or places, count as 1 issue. System categories below include controls related to the system	HVAC system integration (EMCS/BAS)	Chilled water plant and distribution system	Heating water plant and distribution system	Domestic hot water	Heat pump system	Air handling and distribution: Overhead system	Air handling and distribution: Underfloor system	Terminal units	Thermal energy storage	Radiant heating	Radiant cooling	Passive heating/cooling
23		m2	32000	4	Y	Y	0			Y			0			Y
24		m2	13702	2	Y											
25		m2	17000	3			Y			Y						
26		m2	12000	5	Y					Y						
27		m2	12500	4			Y			Y				Y		
28		m2	10000	4	Y					Y				Y		
29		ft2	11400	1						Y						
30		ft2	200460	1												
31		ft2	125000	1												
32		ft2	589140	0												
33		ft2	255490	0												
34		ft2	180316	0												
35		ft2	177838	0												
36		ft2	130844	0												
37		ft2	158979	4				Y		Y		Y				
38		ft2	165030	0												
39		ft2	97920	0												
40		ft2	113699	0												
41		ft2	114666	0												
42		ft2	192001	0												
43		ft2	258600	0												

Project Ref Number	TECHNICAL INFORMATION CONTINUED						COST DATA CONTINUED - Give costs in year of original data; do not correct for inflation	Commissioning costs	Cost unit (currency or labor hours)	Total Cx costs	Tasks performed - Select "Y" if task was performed	Document owner's project requirements
	Natural ventilation or mixed-mode ventilation	Lighting/ daylighting and lighting controls	Electrical	Envelope and infiltration	Utility-related (electric, gas, water, emergency power)	Other - Explain in column E	Currency					
1	Y	Y			Y		EUR (euro)		Currency			
2	Y	Y			Y		EUR (euro)		Currency	5000		
3		Y					EUR (euro)		Currency	5000		
4	0	0		Y			EUR (euro)		Labor	150		
5		Y			Y		EUR (euro)		Currency			Y
6	Y	Y			Y		EUR (euro)		Currency			Y
7							CAD (Canadian dollar)		Currency	199300		Y
8							CAD (Canadian dollar)		Currency	90000		
9							JPY (Japanese yen)		Currency	10480000		Y
10							JPY (Japanese yen)		Currency	19800000		
11					Y		JPY (Japanese yen)		Labor	1200		Y
12					Y		JPY (Japanese yen)		Labor	5000		
13							JPY (Japanese yen)		Currency	20000000		
14							JPY (Japanese yen)		Labor	200		
15							EUR (euro)		Currency	185000		
16				0			EUR (euro)		Currency	26000		Y
17							EUR (euro)		Currency	27000		Y
18							EUR (euro)		Currency	2000		
19							EUR (euro)		Currency	15200		y
20						Y	EUR (euro)		Currency	51000		y
21							EUR (euro)		Currency	15200		y
22							EUR (euro)		Currency	22000		y

Project Ref Number	TECHNICAL INFORMATION CONTINUED						COST DATA CONTINUED - Give costs in year of original data; do not correct for inflation				Tasks performed - Select "Y" if task was performed	Document owner's project requirements
	Natural ventilation or mixed-mode ventilation	Lighting/ daylighting and lighting controls	Electrical	Envelope and infiltration	Utility-related (electric, gas, water, emergency power)	Other - Explain in column E	Currency	Commissioning costs	Cost unit (currency or labor hours)	Total Cx costs		
23							EUR (euro)		Currency			y
24							NOK (Norwegian krone)		Currency	33000		
25		Y	Y				NOK (Norwegian krone)		Currency	110000		
26		Y	Y	Y			NOK (Norwegian krone)		Currency	110000		
27		Y					NOK (Norwegian krone)		Currency	110000		
28	Y						NOK (Norwegian krone)		Currency	110000		
29							USD (US dollar)		Currency	3000		
30							USD (US dollar)		Currency			
31							USD (US dollar)		Currency	71693		
32						Y	USD (US dollar)		Currency	180554		Y
33							USD (US dollar)		Currency	77324		
34							USD (US dollar)		Currency	99050		
35							USD (US dollar)		Currency	49625		
36							USD (US dollar)		Currency	27344		
37			Y				USD (US dollar)		Currency	53035		
38							USD (US dollar)		Currency	59054		
39							USD (US dollar)		Currency	41894		
40							USD (US dollar)		Currency	44541		
41							USD (US dollar)		Currency	52314		
42							USD (US dollar)		Currency	66423		
43							USD (US dollar)		Currency	105591		

Project Ref Number	COST DATA CONTINUED - Give costs in year of original data, do not correct for inflation	Develop Commissioning Plan	Utility bill analysis	Bench marking	Trend analysis (for example, EMCS, data logging, etc)	Building energy modeling/ simulation	Use of diagnostic tools and Cx automation techniques. List tools and/or methods used in column E	Document master list of findings	Calculate energy cost savings for findings	Present a findings and recommendations report	Update system documentation after implementation	Implement operations and maintenance (O&M) improvements	Implement capital improvements	Monitor and verify energy savings	Monitor implemented measures for persistence of benefits	Develop systems manual/ recommissioning manual	Final commissioning report	Other - Explain in column E
1			Y	Y	Y			Y	Y	Y		Y	Y	Y	Y	Y	Y	
2			Y	Y			Y	Y	Y	Y								
3			Y	Y			Y	Y	Y	Y								
4				Y	Y	Y				Y								
5			Y	Y	Y	Y		Y	Y	Y		Y		Y				
6			Y	Y	Y		Y	Y	Y	Y					Y		Y	
7		Y	Y	Y	Y		Y	Y	Y	Y	Y	Y	Y	Y	Y			
8		y	Y	Y	Y	Y		Y	Y	Y	Y	Y	Y	Y	Y			
9		Y	Y	Y	Y	Y												
10					Y		Y	Y	Y	Y	Y	Y		Y	Y		Y	
11		Y	Y	Y	Y	Y	Y	Y	Y	Y		Y	Y	Y	Y	Y	Y	
12					Y	Y	Y	Y		Y		Y		Y				
13								Y										
14		Y	Y		Y			Y	Y	Y		Y		Y	Y		Y	
15		Y			Y			Y		Y							Y	
16					Y			Y		Y		Y			Y	Y	Y	
17		Y						Y									Y	
18					Y			y									Y	
19					y			y		y		y					y	
20					y			y		y			y				y	
21					y			y		y							y	
22					y			y		y							y	

Project Ref Number	COST DATA CONTINUED - Give costs in year of original data; do not correct for inflation	Develop Commissioning Plan	Utility bill analysis	Bench marking	Trend analysis (for example, EMCS, data logging, etc)	Building energy modeling/simulation	Use of diagnostic tools and Cx automation techniques. List tools and/or methods used in column E	Document master list of findings	Calculate energy cost savings for findings	Present a findings and recommendations report	Update system documentation after implementation	Implement operations and maintenance (O&M) improvements	Implement capital improvements	Monitor and verify energy savings	Monitor implemented measures for persistence of benefits	Develop systems manual/recommissioning manual	Final commissioning report	Other - Explain in column E
23		y			y			y		y							y	
24		Y				Y		Y	Y	Y		Y				Y	Y	
25		Y		Y					Y	Y				Y		Y	Y	Y
26		Y		Y					Y	Y				Y		Y	Y	
27		Y		Y					Y	Y				Y		Y	Y	
28		Y		Y					Y	Y				Y		Y	Y	
29																		
30					Y	Y		Y	Y	Y			Y	Y			Y	
31		Y	Y	Y	Y			Y	Y	Y		Y					Y	
32		Y	Y	Y	Y			Y	Y	Y	Y	Y				Y	Y	
33																		
34										Y		Y		Y				
35																		
36						Y												
37									Y	Y		Y	Y	Y	Y		Y	
38										Y		Y	Y					
39																		
40												Y						
41										Y								
42																		
43																		

Source Unit cost

Project Ref Number		Electricity Avg cost per unit (in the first year after commissioning)	Unit	Electric demand Average cost per unit (in the first year after commissioning)	Unit	Fuel type	Fuel Avg cost/unit (in the first year after Cx)	Unit	District chilled water Avg cost/unit (in the first year after Cx)	Unit	District hot water Avg cost/unit (in the first year after Cx)	Unit	Water Average cost per unit (in the first year after commissioning)	Unit
1	Project Data	14	EUR/kWh											
2	Project Data	0.155	EUR/kWh			Natural gas	0.01944444	EUR/MJ						
3	Project Data	0.155	EUR/kWh			Natural gas	0.01944444	EUR/MJ						
4	Project Data					Natural gas								
7	Project Data	0.0526	CAD/kWh	11.85	CAD/kW	Natural gas	0.48	CAD/m3						
8	Project Data	0.074	CAD/kWh			Natural gas	0.54	CAD/m3						
9	Project Data	9.98	JPY/kWh	1774.5	JPY/kW								0.157	JPY/Liters
	Project Data	8.99	JPY/kWh	1517	JPY/kW									
10	Comments	on Cx (but 9.890 in summer daylight & 4.500 at night)		Average in 2005										
24	Project Data	0.7	NOK/kWh								0.7	NOK/kWh		
26	Project Data	0.078	USD/kWh	0.121	USD/kW	Natural gas								
30	Project Data	0.55	NOK/kWh	460	NOK/kW								15	NOK/O her
	Comments												Incoming + outlet water	
31	Project Data	0.55	NOK/kWh	460	NOK/kW	Fuel oil	0.5	NOK/Therms					15	NOK/O her
	Comments					The cost is per kWh							Incoming + outlet water	
32	Project Data	0.55	NOK/kWh	460	NOK/kW	Fuel oil	0.5	NOK/Therms					15	NOK/O her
	Comments												Incoming + outlet water	
33	Project Data	0.02788	USD/kWh						4.67	#N/A	4.75	USD/MMBTU		
34	Project Data	0.02788	USD/kWh						4.67	#N/A	4.75	USD/MMBTU		
35	Project Data	0.02788	USD/kWh						4.67	#N/A	4.75	USD/MMBTU		
36	Project Data	0.02788	USD/kWh						4.67	#N/A	4.75	USD/MMBTU		
38	Project Data	0.02788	USD/kWh						4.67	#N/A	4.75	USD/MMBTU		
39	Project Data	0.02788	USD/kWh						4.67	#N/A	4.75	USD/MMBTU		
40	Project Data	0.02788	USD/kWh						4.67	#N/A	4.75	USD/MMBTU		
41	Project Data	0.02788	USD/kWh						4.67	#N/A	4.75	USD/MMBTU		
42	Project Data	0.02788	USD/kWh						4.67	#N/A	4.75	USD/MMBTU		
43	Project Data	0.02788	USD/kWh						4.67	#N/A	4.75	USD/MMBTU		

Project Ref Number		Electricity — Savings from implemented measures (kWh)	Electric demand — Savings from implemented measures (kW)	Fuel — Savings from implemented measures (m3)	District chilled water — Savings from implemented measures (MJ)	District hot water — Savings from implemented measures (MMBTU)	Water — Savings from implemented measures (liters)
1	Project Data	8310		73443			
7	Project Data	4047000	949	178800			
8	Project Data	60622	164	59898			
9	Project Data	283410					
	Project Data	22100	41				
10	Comments	468500-4464000	1895-1854				
12	Project Data	29570					
13	Project Data		30				
14	Project Data	113000		1800	247000	1424000	2600000
24	Project Data					142382	
26	Project Data	334167					
	Project Data	126523					
30	Comments	In 2004, there is no proof that all the suggested measures have been implemented					
31	Project Data	173288					
32	Project Data	67059					
	Project Data	1059000			6232	4642	
33	Comments	Based on 1st year after RCx			after RCx	Based on 1st year after RCx	
	Project Data	1193000			11779	5006	
34	Comments	Based on 1st year after RCx			MMBTU - Based on 1st year after RCx	Based on 1st year after RCx	
35	Project Data	183000			10155	15064	
36	Project Data	369000			7070	4293	
37	Project Data	2545			781	-244	
38	Project Data	53000			24407	34289	
39	Project Data	339000			9787	1399	
40	Project Data	35000			14927	11662	
41	Project Data	220000			17777	2682	
42	Project Data	145000			6866	2517	
43	Project Data	740000			24087	6046000	

APPENDIX B: The State of Building Systems Commissioning

The following sections on the Asia Pacific, Europe, and North America regions provide an overview of the state of building systems commissioning. Each national assessment aims to present the national drivers, leading efforts and an indication of their success (if known). Fifteen countries are represented in this review.

Asia Pacific

In the Asia Pacific region, the construction industry is experiencing unparalleled growth. China alone reports the annual construction of two billion m^2s of floor space, (Qiu, 2008). The environmental impact of this growth has global repercussions. It is therefore critical that the building industry adopt building commissioning as a standard practice to improve the quality and efficiency of design, construction, and operations.

Asia Pacific countries are leveraging national energy laws, institutional mandates, and non-profit organizations to foster development of building systems commissioning. For several countries in the region, recent exposure through international symposia and conferences sparked interest that has developed into national research projects. This exchange of information is particularly useful for countries that have natural synergies, such as climate, that facilitate transfer of technologies or guidelines. Through these cooperative exchanges, countries are able to leapfrog technologies and accelerate adoption.

Australia: Indoor environmental quality assessments, rising energy costs, and a greater understanding of the link between comfort and productivity are driving the demand for improved energy efficiency of buildings in Australia. At present, most of the commissioning projects are associated with major contracts and do not address smaller projects. Furthermore, the fully integrated approaches to project commissioning, promoted by the Chartered Institution of Building Services Engineers (CIBSE) and American Society of Heating, Refrigerating and Air-conditioning Engineers (ASHRAE) commissioning guidelines, are not well understood by the commercial sector, and this educational need is yet to be addressed (EcoLibrium 2006).

Awareness of the importance of the commissioning process is increasing, largely due to the market influence of voluntary, performance-based rating schemes such as the National Australian Built Environment Rating System and the Green Building Council's Green Star program. Through this mechanism, several large commercial real estate owners and governmental agencies have made commitments to achieve a certain rating in their building stock. The implementation and promotion of these rating schemes have the potential to trigger the demand for energy audits and commissioning in the existing building stock. Deakin University is developing rapid reporting techniques to facilitate building performance improvement in the commissioning process (Nakahara et al. 2007).

China: Mainland China has seen several construction booms. In the 1980's the rapid growth of residential construction prompted the Ministry of Construction to develop energy-efficient building codes. These codes did not include commissioning and quality, in general, was considered poor. In the 1990s, testing and commissioning concepts diffused into the Chinese buildings industry as part of the western management techniques from numerous overseas, joint venture construction projects (Chow et al. 2006). International partnerships now also extend to academic research. Chinese universities, including Tsinghua University, are conducting research on building optimization, building commissioning and other related work. In 2008, China's Ministry for Construction released acceptance codes for building services equipment and building systems, though no information could be obtained regarding the administration and enforcement of these requirements.

Hong Kong was one of the first adopters and developers of the building commissioning process due to early exposure to the United Kingdom's commissioning model. As early as 1990, the Hong Kong Government published twelve booklets based on the CIBSE Commissioning Codes. The documents were developed into commissioning specifications released in 2002 and are intended to be incorporated into government building project contracts (Chow et al. 2006). Furthermore, in 2004, Hong Kong introduced a new voluntary scheme, termed the Consolidated Environmental Performance Assessment Scheme (CEPAS), to promote the design of environmentally-friendly buildings. CEPAS incorporates building commissioning as a major element for performance assessment.

The Hong Kong Building Commissioning Centre (HKBCxC) was established in December 2004 as a non-profit organization whose mission is to promote the establishment of a standardized approach to building commissioning in Hong Kong. The HKBCxC organizes programs for continuing professional development, certification services and publishes guidebooks on the management of building commissioning, field measurements, and system tuning. In the "Practical Guide to Building Commissioning Management", Chow et al (2006) report that independent contracts for commissioning and retro-commissioning are soaring in Hong Kong.

Japan: In 2006, building energy performance reporting became mandatory under the Energy Conservation Law. The reports are based on simple performance tests of the components and systems that have the largest impact on the energy consumption of heating, ventilation, and air-conditioning systems. In practice, there is significant variation in the implementation approach because no standard test procedures are specified and it is unclear whether relevant problems could be investigated adequately through the data contained in the reports. Hence, aspects of the commissioning process are drawing more attention in Japan's building sector. In existing buildings, various approaches for retro-commissioning are commonly implemented, but initial commissioning for new construction is not common.

In 2005 the Society of Heating, Air-Conditioning and Sanitary Engineers of Japan (SHASE) technical committee on commissioning issued a guideline on the building services commissioning process. The Building Services Commissioning Association (BSCA), a non-profit organization launched in 2004, provides seminars about commissioning technologies in major cities and has undertaken cooperative activities with Asian countries such as China (including Hong Kong), Taiwan, and Korea. It also continues to compile commissioning documentation and tools through actual commissioning projects and research. BSCA's strategy is to establish a certification program for commissioning engineers, including the Commissioning Authority, and to educate the construction industry and related government sectors.

Energy policy is playing an important role. The Ministry of Economy, Trade and Industries is interested in a new business model, based on building commissioning, to enhance energy efficiency of new and existing buildings and The Ministry of Land, Infrastructure and Transportation is promoting use of lifecycle energy management with a newly developed simulation tool. Market demand for commissioning is believed to be strong but mandates, based on energy and environmental policy, are needed for building owners to apply building commissioning to new construction.

Taiwan: In Taiwan, building commissioning for new construction is considered established practice. The Taiwanese government has issued a mandate requiring that all public projects with a budget over 15 million U.S. dollars acquire the Green Building Label before a construction permit can be granted. Testing, adjusting and balancing (TAB), commissioning, and the use of a building energy management system (BEMS) are prerequisites. However, the process implemented in Taiwan does not represent the full range of actions from design to operations. The commissioning role in Taiwan for existing buildings is typically similar to that served by energy service companies in Japan and the USA. Taiwan is involved in international activities to expand national practices as deemed necessary. For example, a review of the Green Building Label (similar to Leadership in Energy and Environmental Design (LEED) for new construction in the USA and CASBEE in Japan) suggested that a commissioning plan be added.

A national-scale project was launched in 2003, by the Architecture and Building Research Institute, Ministry of the Interior, to renovate all the central heating, ventilation, and air-conditioning (HVAC) systems in governmental buildings for energy conservation. The renovation process includes system diagnostics, remedial strategies, establishing engineering jobs, contracting, TAB, commissioning, and system performance validation through the BEMS. To date, 22 million USD have been spent with an overall energy-savings of 22 % and an average payback of five years. The success has since led to a series of demonstration projects for civil and governmental buildings. In 2008, Taiwan will launch another five-year program where the energy savings effort will be enhanced with greater system fine-tuning and commissioning as a means to support the Kyoto Protocol and global CO_2 emission reduction efforts.

Korea: The Korean Institute for Energy Research (KIER) is working to promote the commissioning process through energy conservation and quality assurance measures for new construction. Commissioning projects have been carried out by KIER as part of their research work, and have also been implemented in several buildings financed by foreigners. However, there is no recognized national standard and the Korean government has no intention to mandate the commissioning process.

Europe: The countries of the European region present significant differences in their building delivery processes as well as their emphasis on energy efficiency and measures for quality assurance. For most, with the exception of the UK, the commissioning process is quite new. However, the European Commission established the European Performance of Buildings Directive, EPBD (EC 2002), to promote the improvement of energy efficiency and building performance. Four requirements to be implemented by the Member States are to: 1) develop a framework for a methodology to calculate the integrated performance of buildings, 2) set minimum standards in new and existing buildings, 3) certify the energy performance of buildings, and 4) inspect and assess heating and cooling installations.

According to reports from the member states, the EPBD poses significant challenges in terms of its practical implementation, including difficulties associated with the transfer of requirements into existing building practices under a range of climates. However, because the commissioning process is well aligned with the goals of the EPBD, several national research programs are introducing commissioning tools as a means to address the requirements of the directive. In many countries, commissioning tasks are focused on the building handover, or performed as part of the facilities management. However, for commissioning to have a real impact on savings, the review must begin at the pre-design phase, where changes are easier and more cost-effective to make. It is anticipated that the increased attention to energy efficiency in buildings will lead to greater application and consistency of commissioning through the building lifecycle.

Belgium: Commissioning research has been underway in Belgium for several years, and mainstream awareness of the importance of commissioning of low-energy buildings has increased due to the implementation of the EPBD and the introduction of the passive house concept for very low energy building. Energy performance laws, which set requirements for the energy performance and indoor climate for most buildings requiring a building permit, exist in the Flemish Region and are under development in the Brussels and Walloon regions.

Czech Republic: Building commissioning is a new concept in the Czech Republic. Only some aspects of building commissioning are implemented as a part of the facility management and energy auditing processes that are related to EPBD implementation. Under the IEA Annex 47 project, researchers have developed tools that support additional aspects of the broader commissioning process, including new control system energy services.

Finland: Historically, individual contractors and builders in Finland have managed commissioning-related activities as part of their quality assurance measures. More recently, emphasis has been placed on the development and implementation of "energy auditing" procedures. In 2002, a Finnish national research program called CUBE was launched to improve the performance of building services. This program includes a national R&D project to develop Finnish procedures for building commissioning, focusing on the indoor air quality and energy efficiency of buildings. The Finnish term and concept of "toimivuuden varmistaminen (ToVa)," an adaptation of building systems commissioning, is being promoted. Practical testing and further development of the guidebook and tools are underway. Methods and tools to support the commissioning of buildings and their subsystems throughout the phases of the building life-cycle are being developed.

France: The commissioning process is just beginning to take hold in France and a national guideline on commissioning is under development. In practice, commissioning is implemented in the operations phase, though there is greater interest from large building owners for more complete commissioning plans as a result of ongoing national research and new requirements under the EPBD. Current research is focused on automating the commissioning process to improve performance, pushing for early implementation of the commissioning process from design through certification, and developing tools and procedures for specific building applications (e.g., schools).

Germany: Commissioning for new construction is not established or even required as a third party service in Germany. German law (HOAI) dictates that architects and design engineers perform the following tasks within the construction administration and construction supervision:

- supervising acceptance and performance tests and statement of deficiencies;

- collecting/compiling and delivering as-built documentation, operating manuals, and acceptance protocols; and

- supervising the rectification of deficiencies that fall under the two-year contractor's warranty period or the five-year design team warranty period from date of acceptance.

This approach, initiated in 1976, presents a model for internal commissioning during the construction phase and, in conjunction with energy conservation laws, has resulted in higher performing buildings. Efforts to implement the full commissioning process beginning in pre-design have been more recent. In one German national research program on energy optimization in buildings (ENOB), more than twenty demonstration buildings surpassed national energy consumption standards by 50 % without incurring additional building costs. The program has been extended and now supports the design process, commissioning, and monitoring of the first two years of operation. ENOB also supports several projects focusing on improved commissioning and operation of innovative buildings. Energy agencies and utilities support energy efficiency through contracting and public private partnerships. Two other projects, ModBen and Building EQ, deal with performance evaluation of existing buildings.

The Netherlands: In the Netherlands, practitioners agree on the importance of building commissioning. However, it is currently only implemented in the installation and formal handover of selected buildings. A pilot project on functional performance tests conducted in forty buildings identified that 70 % of the systems tested were malfunctioning, leading to increased energy use and reduced comfort. To improve system operation, large-building owners are investigating performance contracting for operations and maintenance, based on well-defined criteria. The commissioning process plays an important role. The Netherlands Organization for Applied Scientific Research is involved in the development of national standards concerning the energy performance of new buildings to promote the implementation of the whole commissioning process. **Norway:** Building commissioning is not an established practice in Norway. A draft national commissioning guideline was developed in 2007 to promote the life-long commissioning of building HVAC systems. The guideline is currently under review and several large governmental and private building owners are involved in the effort to verify, document and implement suitable tools to provide continuous control of energy and indoor environment during the life of the building.

United Kingdom: The UK developed the earliest commissioning codes and provided the basis for similar work in many other countries. CIBSE published the first commissioning code on air-distribution systems in 1960 and subsequently released codes for other types of equipment. The emphasis of the commissioning codes was originally post-construction commissioning. Bordass (2008) reports that independent commissioning engineers were usually appointed as part of the design and build team in the late 1970's and early 1980's. However, as markets became more competitive, in the late 1980's and 1990's, commissioning was less commonly sold as a separate service and eventually became a subcontractor role, which had a negative impact on quality due to cost-cutting pressures. Today, initial commissioning has become routine for large projects and quality has once again improved but there are still major challenges in that the commissioning period is often squeezed when other delays impact the delivery date. The current set of CIBSE commissioning codes includes: air distribution systems, boilers, automatic controls, lighting, management (Code M), refrigeration, and water distribution systems. Code M shows an important change in the approach to the initial commissioning process by stating that the commissioning manager should be appointed early in the design phase in order for the system to be designed as commissionable.

Other organizations have also been directly involved in improving the commissioning industry. The Building Services Research and Information Association (BSRIA), a non-profit consulting design, construction, and operations organization is leading the "Soft landings" development with support from the Usable Buildings Trust, a non-profit organization dedicated to improving building performance through the better use of feedback. The Soft landings procedure aims to reduce the loss of information by extending the service of the design and building (and commissioning) team to facilitate fine-tuning and debugging at building handover in order to achieve a closer match between design targets and building operation, and to allow individuals to learn from the experience (BSRIA 2008). The Commissioning Specialists Association (CSA) is focused on career development, including training and certification of individuals.

Although retrocommissioning is not widespread, there is expected to be a major increase in its implementation as demand for improved energy and carbon performance increases, driven by new requirements of the EPBD, occupants, and building owners.**North America**

In North America, the concept of building commissioning began with the Code of Practice for Commissioning Mechanical Systems in Buildings that was developed in 1986 by the Standing Committee of Consulting Engineers and Mechanical Contractors of British Columbia. Today there are several, industry-recognized guidelines on the commissioning process: ASHRAE Guideline 1-1989 The HVAC Commissioning Process (revised in 1996), and ASHRAE Guideline 0-2005, The Commissioning Process. Although building systems commissioning is established practice in both Canada and the USA, the process is not widespread. Many of the existing resources are focused on conventional HVAC systems and there is need for information on other types of systems, particularly due to increased interest in non-conventional, low energy systems.

In Canada and the USA, awareness has increased through professional organizations, certification programs (e.g., LEED), large-owner mandates, and energy-efficiency initiatives. However, market barriers are significant. There is a need for greater awareness of the benefits and cost of commissioning to increase demand, and a need for more training and certification of commissioning providers to increase the supply. At present, building owners lack the access to experience and lessons learned in easily accessible, creditable and persuasive formats that would facilitate their investment decisions. Furthermore, tools and standardization are needed to reduce the cost of commissioning and improve the cost-benefit ratio for greater uptake. This includes automated tools, data on cost-benefits, and clear specifications for key building performance metrics, monitoring methods, and energy calculation methods. Details of national initiatives follow.

Canada: Existing building commissioning in Canada is at the early stage of its development. While the demand for commissioning services remains stagnant, there are encouraging signs of activity transformation favorable to the creation of a better structured and more efficient market place. As more and more building owners and managers strive for energy conservation and building sustainability, their first step consists to improve their building operation efficiency. With the increase of energy prices, low cost technology availability and their practical advantages have strengthened the demand for energy-saving HVAC controls.

In order to support institutional and commercial businesses in this shifting market, the Ministry of Natural Resources Canada and its energy science and technology organization, CanmetENERGY, provides knowledge, services and tools for recommissioning, commissioning and ongoing commissioning projects. They offer to building engineers, technicians, owners, managers and other similar stakeholders a suite of capacity-building and decision-making tools, including guides, training material, awareness seminars, pre-screening and benchmarking tools and case studies. CanmetENERGY developed an application software tool (DABO™) to continuously monitor, detect and diagnose the operation inconsistencies and poor performances of a building's electromechanical systems.

Furthermore, the Demand Side Management Working Group (DSMWG) created by Natural Resources Canada collaborates with utilities, public works, green building associations and standard organizations on energy management to accelerate the training and certification of commissioning providers and to raise stakeholders' awareness on the benefits of commissioning work. The persistence of energy savings has also attracted the interest of DSMWG and action has been undertaken to implement a new Energy Management Information System (EMIS) component to the recommissioning process.

It is also worth mentioning that the Canadian Standard Association is preparing a new standard on commissioning that should come into effect in 2011.

USA: Rising energy costs and a shift in public policy that emphasizes the need for energy independence are driving energy conservation. In the absence of national commissioning requirements, individual state codes and institutional mandates have proven to be strong market drivers, though enforcement mechanisms have been problematic.

The buildings industry is working to address market barriers and improve the quality of its services. Non-profit organizations, including ASHRAE, Portland Energy Conservation, Inc. (PECI), the California Commissioning Collaborative and the Building Commissioning Association (BCA), provide access to industry resources for both providers and building owners, including:

A library of published papers and commissioning guides for retro-, initial, and on-going commissioning; tools: design guides, operations and maintenance best practices, case studies database; Sample documents: commissioning plans, specifications, functional tests, checklists; and training and certification programs, career and provider directories.

In recent years, market indicators show growth in the number of firms offering commissioning services. In 2008, 267 commissioning provider firms and 665 members registered with the BCA. There has also been a dramatic increase in commissioning certification and training provided by professional organizations, which indicate a demand by engineering professionals for more knowledge.

Utilities and government agencies have also invested resources on research and technology demonstration activities in an effort to stimulate the market for commissioning. More investment is needed to significantly improve the energy efficiency of the existing building stock, to meet the low energy targets for new construction, and to address the projected shortage of skilled and certified providers.

APPENDIX C: Relevant Papers and Presentations

Claridge, D.E., "Methodologies for Determining Persistence of Commissioning Benefits," Proc. of 7th International Conference for Enhanced Building Operations – Maximizing Building Energy Efficiency and Comfort Permission to reprint granted on 12/1/11.

 Toole, C. and Claridge, D.E., "Review on Persistence of Commissioning Benefits in New and Existing Buildings", Proc. of 7th International Conference for Enhanced Building Operations – Maximizing Building Energy Efficiency and Comfort, Part II, Paper VI-4-4, Shenzhen, China, November 6-8, 2006, 15 pp. Permission to reprint granted on 12/1/11.

Lin, G. and Claridge, D.E., "Retrospective Testing Of An Automated Building Commissioning Analysis Tool (ABCAT)," Proceedings of the 3rd International Conference On Energy Sustainability ASME, July 19-23, 2009, San Francisco, CA, USA. Copyrighted by ASME. Permission to reprint granted on 12/1/11.

Insert PDF files of 3 papers and check page numbering.

Appendix D: Annex 47 Participants

Name	Country	Affiliation
Alexis Versele	Belgium	KaHo St-Lieven
Hilde Breesch	Belgium	KaHo St-Lieven
Stephane Bertagnolio	Belgium	University of Liege
Daniel Choinière	Canada	Natural Resources Canada
Zhongxiian Gu	China	TNO Beijing
Yongning Zhang	China	Tsinghua University
Karel Kabele	Czech Republic	Czech Technical University
Pavla Dvorakova	Czech Republic	Czech Technical University
Michal Kabrhel	Czech Republic	Czech Technical University
Mika Violle	Finland	Helsinki University of Technology
Jorma Pietilainen	Finland	Technical Research Centre of Finland (VTT)
Hannu Keranen	Finland	Helsinki University of Technology
Lari Eskola	Finland	Helsinki University of Technology
Satu Paiho	Finland	Technical Research Centre of Finland (VTT)
Hossein Vaezi-Nejad	France	Dalkia
Oliver Baumann	Germany	Ebert & Baumann Consulting Engineers
Steffen Plesser	Germany	Institute of Building Services and Energy Design (IGS)
Christian Neumann	Germany	Fraunhofer Institute for Solar Energy Systems
Dirk Jacob	Germany	Fraunhofer Institute for Solar Energy Systems
Anatoli Hein	Germany	Institute of Building Services and Energy Design (IGS)
Michele Liziero	Germany/Italy	Politecnico di Milano, Guest Scientist ISE
Jochen Schaefer	Germany	Ebert & Baumann Consulting Engineers
Shengwei Wang	HK/China	Hong Kong Polytechnic University
Xinhua Xu	HK/China	Hong Kong Polytechnic University
Zhenjun Ma	HK/China	Hong Kong Polytechnic University
Zhou Cxiang	HK/China	Hong Kong Polytechnic University
Xiao Fu Linda	HK/China	Hong Kong Polytechnic University
Na Zhu	Hong Kong	Hong Kong Polytechnic University
Zoltan Magyar	Hungary	University of Pecs
Csaba Fodor	Hungary	University of Pecs

Name	Country	Affiliation
Harunori Yoshida	Japan	Kyoto University
Motoi Yamaha	Japan	Chubu University
Mingjie Zheng	Japan	Sanyo Air Conditioning
Yasunori Akashi	Japan	Kyushu University
Hiroo Sakai	Japan	Hitachi Plant Technologies
Katuhiro Kamitani	Japan	Tonets Corporation
Ryota Kuzuki	Japan	Tokyo Gas Co
Katsuhiko Shibata	Japan	Takasago Thermal Eng. Co
Fulin Wang	Japan	Kyoto University
Masato Miyata	Japan	Kyoto University (student)
Hirotake Shingu	Japan	Kyoto University (student)
Hiromasa Yamaguchi	Japan	Kansai Electric Power Co
Ryusi Yanagihara	Japan	Tokyo Electric Power Co
Hideki Yuzawa	Japan	Nikken Sekkei Research Institute
Takao Odajima	Japan	Takenaka Corp.
hirobumi ueda	Japan	Osaka Gas CO., Ltd
Masahiro Shinozaki	Japan	Kyushu Electric Power Co.
Katsuhiro Kamitani	Japan	Tonets Corporation
Mingjie zheng	Japan	SANKO AIR CONDITIONING CO.,LTD
Katsuhiko Shibata	Japan	Takasago Thermal Engineering Co..Ltd
Vojislav Novakovic	Norway	Norwegian University of Science and Technology (NTNU)
Natasa Djuric	Norway	NTNU
Marko Masic	Norway	NTNU
Vojislav Novakovic	Norway	NTNU
Henk Peitsman	the Netherlands	Netherlands Organization for Applied Scientific Research (TNO)
Luc Soethout	the Netherlands	TNO
Ipek Gursel	the Netherlands	University of Delft
Natascha Milesi Ferretti	USA	National Institute of Standards & Technology
David Claridge	USA	Texas A&M University
Hannah Friedman	USA	Portland Energy Conservation Inc.

Name	Country	Affiliation
Omer Akin	USA	Carnegie Mellon University
Ashish Singhal	USA	Johnson Controls
Tudy Haasl	USA	Portland Energy Conservation Inc
Phil Haves	USA	Lawrence Berkley National Laboratory